CITIES OF GOLD

History and Tales of the Cripple Creek-Victor Mining District

By Brian H. Levine

CENTURY ONE PRESS

2325 East Platte Avenue
Colorado Springs, Colorado 80909

Library of Congress Catalogue
Card number 81-50063
ISBN 0-937080-09-8
Composed, printed, and bound in U.S.A.

Typesetting and layout by
Prepress Advertising
Denver, Colorado

Special regards and gratitude . . .

Wayne McCormick (President of the Chicago-Victor Gold Mining Company,

Alice Ashley (Proprietor of Ye Olde Bottle Shop, Victor, Colorado)

Ed Hunter (Head Mining Engineer for Texasgulf in the Cripple Creek-Victor Mining District)

Dayton Lummis (Curator of the Cripple Creek Museum)

Photo Credits:

Western History Department Penrose Public Library
Colorado Springs, Colorado
Denver Public Library Western History Department

Dedication . . . .

To Myron and Jacquelyn — love

Cities of Gold

As a carpenter, Stratton helped build Colorado Springs. As the owner of the Independence Mine, he "ruled" Little London.

By Brian H. Levine

INTRODUCTION

Riches still flow out of the Colorado Rockies. Gold! The prospects could revive the towns of Victor, Goldfield, and Cripple Creek as bustling mining cities. To date, some $800 million in gold in more than 22 million Troy ounces have been taken from the Cripple Creek-Victor Mining District. Geological speculations insist there is more to be discovered — conservatively, 80 million more ounces. It is still possible literally to dig one's fortune out of the hillsides of the area.

When Winfield Scott Stratton was mining his Independence strike, taking out strictly $120,000 worth of ore a month, the price of gold per Troy ounce was only $20.67. Today, the value of gold has increased to an average of $600 per ounce. Stratton was quoted as saying, "Gold is worth more in the ground than out." In less than eighty years since Stratton's death, gold has increased its value by three thousand percent. The profits could soon be as magnificent as they once were. And the market is wide open. Like a grand bank account totaling in the billions and accruing vast amounts of interest daily, the granite that encases the gold of the Cripple Creek-Victor Mining District has made Stratton's statement a proverb.

But where does the prospective gold-seeker begin? What does he look for in the rock formations? What are the chances of finding gold in the area? Is there a possibility of a gold rush similar to the one that occurred in the region during the 1890s?

A brief tour of the mining district's history gives the modern-day prospector an excellent view of the locations of the major gold mines, the lives of the people who were attracted by the gold and the growth era of the most baffling mining district in the world.

For almost fifteen years the people of Colorado Springs told the persistent prospector Bob Womack he was crazy. Everyone of "Little London" (Colorado Springs) knew there was no gold in the hills behind Pike's Peak. Womack's claim had to be absurd. They called him Crazy Bob.

In 1890 word spread that tenderfeet were mining gold out of the Cripple Creek region with farm ploughs and pitchforks.

ENTERING INTO THE DISTRICT

There are many scenic routes into the region from the Eastern Slope. Two roads, one through Phantom Canyon and the other over Shelf Road, lead up to the district from Canon City. U.S. 24 (Ute Pass) out of Manitou Springs, and the subsequent cut-off of Highway 67 from Divide, is the most traveled route. Gold Camp Road through Cheyenne Canyon is the direct

Womack prospected in the area behind Pikes Peak for over a decade. After he discovered what he had been searching for, he sold his dream for $500.

road out of Colorado Springs. All the roads offer awesome mountain scenery. U.S. 24, through Woodland Park and Divide, is the most expedient route to Cripple Creek. By far, however, Gold Camp Road is the more adventuresome. A smoothly graded forty-five mile dirt road over the old Short Line Railway foundation inspires the pioneer spirit in everyone.

Gold Camp Road rises from Colorado Springs within a mile of the regal Broadmoor Hotel. Almost immediately the traveler is filled with visions of history. A one-way tunnel hewn through pink Pikes Peak granite beckons. Passing through this first of many one-way tunnels, the view of the Spanish Peaks and Raton Pass is lavishly framed in the distance. Driving a short way up this road, the picturesque view of Colorado Springs on the plains far below unfolds. Vice-President Theodore Roosevelt emitted a powerful "Bully!" from his luxurious club car as he rode up this same pass on the Short Line Railway into the district in 1901.

Started by Jimmie Burns, part owner of the fabulously rich Portland Mine, and planned out by Irving Howbert and A.E. Carlton, the Short Line Railroad was an instant success. Tourists from all over the world were ecstatic over the original forty-five mile journey up Cheyenne Canyon. With its variety of club, passenger, flat and box cars, the Short Line began a price war with its competitor, the Midland Terminal which ran from Colorado Springs to Divide on what is now U.S. 24 and Highway 67. During the competition, freight charges dropped to $10 a ton for gold ore, and the rates for passengers dropped from $2.25 to 25 cents — then were free for a brief time.

Continuing through the majestic scenery of Gold Camp Road, the trail intersects with High Road forty miles into the journey. Here the goldseeker has a choice of routes. All lead back into history. The original Short Line Railway went on to Cameron on a northwesterly route. Cameron, platted by the Woods Brothers, sported a recreational area for the miners and their families called Pinnacle Park, with dance halls, a baseball field, and a small zoo. Close to this point, the Short Line met its two electric sections, splitting off to Victor one way and Cripple Creek the other. On the northwesterly course, the prospector would be well on his way to the town of Cripple Creek. Keeping to the southwesterly route,one of the richest pieces of real estate in the world is only a short distance away.

Gold Camp Road is what is left of the Scenic Short Line Railroad which so impressed Teddy Roosevelt.

Part of the Cathedral Bluffs which luxuriously decorate Gold Camp Road.

Just before entering the town of Goldfield, the remains of the most astonishing gold camp in the world start to bespeckle the surrounding hillsides. Gold Camp Road suddenly turns into an outdoor museum with a multitude of different paths to explore and sights to encounter. At its highest point, Gold Camp Road, trailing over Victor Pass, reaches an elevation of 10,200 feet above sea level. But the loftiest point of the entire mining district is the ghost town of Altman atop Bull Hill at 10,700 feet.

In terms of gold ore value, the town of Victor was the richest portion of the district. More specifically, Battle Mountain, rising out of Victor, possessed the granite strata which yielded the greatest amount of high-grade ore in the area. Three of the most profitable mines — the Independence, Portland, and Strong — produced together upwards of $125 million in gold at the price of $20.67 per ounce. This spectacularly wealthy mountain jutts out of the already high plateau of the district and separates the town of Goldfield from Victor.

Goldfield was platted as a township in 1895. Founded by the owners of the Portland Mine, the town filled its boundaries as the production in the Portland became tremendous. In 1894, the "camp" of Goldfield was part of Western Federation of Miners fight for Unionism. The spirit of the union miners had led them to blow up the Strong Mine before the state militia could curtail their activity. Despite its destructive actions, Goldfield became the third largest town in the district and had its civic eye upon unifying itself with Victor. But the town could not compete with the business sectors of Victor and Cripple Creek. It became a largely residential community serving the miners and their families who worked in the mines of Battle Mountain.

After Gold Camp Road meets Highway 67, the town of Victor appears. It is evident that Victor was the mining center of the district while Cripple Creek was its financial and social core. Four mines — the Gold Coin, Strong, Ajax, and Dillon — are still within Victor's city limits. When the Woods Brothers, in 1896, planned out and began to excavate a site on Diamond Avenue for their Victor Hotel, a curious assayer wandered by and noticed that the dirt pushed up into mounds beside the large pit contained rich ore. The plans for the hotel were unhesitatingly scrapped. A shaft was sunk at the site, and within a year the Gold Coin Mine had produced $90,000 worth of gold. By 1899, the mine was paying out $210,000 in dividends to its stockholders.

Even today, the town of Victor is a living picture of Old West romance — exquisite material for intriguing mythology.

Victor now rests on two miles of underground mine drifts. A 3,700-foot tunnel runs under the town and Squaw Mountain from the first level of the Gold Coin to what was once the enormous Economic Extraction Mill. The chlorination mill is in ruins today, but the miles of mining tunnels under Victor remain.

CRIPPLE CREEK-VICTOR RIVALRY BEGINS

Victor, Colorado, has long since been placed in the shadows of Cripple Creek. Although gold was originally discovered by Crazy Bob Womack in Poverty Gulch, southeast of Cripple Creek, and the district was formerly named after the crooked stream that had so often crippled grazing cattle stumbling upon its banks, it has been unjust to the town of Victor to have given it second billing for so long.

First staked out in the fall of 1891, Victor was originally the town of Lawrence — 160 acres plotted out for the Mount Rosa Mining, Milling and Land Company Placer. This was three months after Winfield Scott Stratton discovered his Independence Mine through a "vision." Lawrence had been founded to serve the prospectors in the Squaw and Battle Mountain area. As the wealth of the discoveries of the Independence, Portland, Ajax, Strong, and other mines grew, the townsite of Lawrence became Victor when the Woods Brothers purchased the area from J.R. McKinnie for $1,000 in 1893.

Victor, by 1896, had become the second largest settlement in the district through three major strokes of fortune — the wealth of Battle Mountain, the Woods family empire, and the railroads. Because of this fabulous luck, the town of Victor became known as the "City of Mines."

The riches of Battle Mountain had been the greatest of all Victor's astounding fortune. After Stratton's "vision" of marvelous wealth had come to him on the night of July 3, 1891, Battle Mountain was never the same.

For years, Stratton had heard Crazy Bob Womack drunkenly roll off tales of the gold ore that he had found up at a cow pasture homesteaded by Womack's family. Every time Womack came down from the hills behind Pike's Peak into Colorado Springs he had a new piece of "float" with him and another wild story about gold. The year 1891 was no different. Womack had been telling his tales for more than ten years. In the past, everyone had believed Bob to be nothing more than a likeable drunk. But this time Crazy Bob's stories attracted considerably more interest.

Few in Little London ever gave thought to the reasons why Bob Womack believed there was gold in the hills behind Pike's Peak. Had they believed, maybe they wouldn't have been left out of the first rush.

Prospectors had been searching through the area since 1865. Levi Welty and his sons settled there in 1865. A rancher and prospector named Requa had been scouring the hillsides in the years before Welty. H.T. Wood found gold ore in the vicinity of Mt. Pisgah in 1874. And no one is entirely sure how long the old prospector, Victor Adams, after whom the Woods Brothers named the town of Victor, had been picking through the rocks of the district.

When the Womack family homesteaded in the area, Bob had a wealth of old prospectors to confer with. He must have learned a great deal about the peculiar form of the gold ore of the area. The ore was not brilliant yellow because it was not oxidized placer nuggets. The ore was found in combination with tellurium and silver. This gave it a silvery appearance rather than a sunny one.

The Womack family ranched in the area for more than a decade then slowly began to sell off their land. The first major portion was sold to a couple of men named Thompson and Anderson. These two in turn sold the land to a Denver real estate firm, Bennett and Myers. Bennett and Myers foreclosed on some of the land but kept a tract which they called the Broken Box Ranch. This ranch was run by George Carr.

Womack and Carr became friends. As Carr tended the Broken Box, Womack continued to search for gold. Bob built a small log cabin in Poverty Gulch and in January of 1891 began to dig a hole which would eventually tap the El Paso lode. The ore that Womack took out of his claim excited some people in Colorado Springs.

When October of 1891 breezed in, Carr was frenzied by the scores of prospectors squatting on the Broken Box. He sent word to Bennett and Myers in Denver. Bennett and Myers decided to sell lots to the prospectors rather than run them off. The town of Fremont was platted.

The first lots went for between $25 and $50. Within a few months, lots would be going for hundreds and then thousands of dollars. While the prospectors of the area debated whether to call the land the Womack

Cripple Creek, circa 1900. Over 80 years the financial center of the district looked-and was-modern.

District, Bennett and Myers were in a race with real estate interests of Colorado Springs to establish the townsite. Names for the town were abundant: Fremont, Hayden Placer, Freeman Placer, Moreland, and Poverty Gulch. Finally the name Cripple Creek was decided upon.

Womack had persuaded the respectable Ed De La Vergne to inspect his Poverty Gulch claim. De La Vergne had been impressed enough to contact Count Pourtales who had been engaged in building the Broadmoor Casino north of Colorado Springs. Pourtales had borrowed $250,000 from his friends in London to build the casino, but he needed more funds. Obviously Count Pourtales liked a fair gamble. Soon Pourtales, after being encouraged by De La Vergne, obtained favorable publicity for Womack and his Poverty Gulch claim. Crazy Bob's bad luck streak of more than fifteen years looked as though it were about to change.

Bennett thought he would never make a cent off the land that Myers had encouraged him to buy simply because the fishing had been good. The "Mount Pisgah hoax" of 1884, which had not taken place at Mount Pisgah at all but at Mount McIntyre some 13 miles west, was still in Bennett's mind. He saw nothing but foolishness in the whole scheme, but when he saw hundreds of prospectors scouting out the land around the Broken Box Ranch he knew there had to be something behind this "gold rush."

Despite Bennett's skepticism, the townsite of Cripple Creek was growing. However, community activity did not get under way in its official construction until November 1891.

By the winter of 1892, prospectors had staked nearly all of the land in the 6.4 square-mile district. But the greatest strikes of all were yet to come. o come.

Publicity and outside interests had encouraged Stratton to do some prospecting of his own. Yet he would not admit it to Womack. Instead, he and W.G. Fernay went up to Beaver Creek near St. Peter's dome to search for cryolite deposits — a fluoride mineral of considerable value used in the production of aluminum. They had no luck, and Stratton and Fernay decided to go up to the Cripple Creek region.

Stratton had not been prepared for the sight. Hundreds of men were pouring over the aspen and pine laden hillsides. There had been so many holes dug in the ground that men were falling into them just as the cattle had crippled themselves in the rugged gulches years earlier. Word reached Stratton and Fernay that the tenderfeet were finding gold all over while the prospectors were still looking for it where it ought to be.

Cripple Creek was like no other gold-mining district. It was the exception to every mining and geological rule. All the seasoned miners of such silver towns as Leadville, Tin Cup, Creede, and Aspen were saying there could not be any valuable minerals in the area. But the tenderfeet — plasterers, carpenters, farmers, druggists — were getting rich by digging through the surface dirt with ploughs, rakes, and any other agricultural tools at hand.

Often, it had been possible to tell the difference between a miner and a tenderfoot simply by the tool he used to dig for ore — pick or plough.

Stratton and Fernay wandered the hills looking for either prospective land or any land that had not yet been plainly staked out. Atop Bull Hill on July 3, 1891, Stratton and Fernay pitched their camp in disappointment. All had been fruitless in the past months. Late that night, however, Stratton had a vision of a peculiar granite formation he had seen a few days before on Battle Mountain. The vision persisted. Seventeen years of failure as a prospector was about to end.

Stratton could not get the vision out of his mind the entire night. When the first rays of the sun splashed upon the pink granite of Pike's Peak, Stratton had packed up his mule and was on his way to stake his claim on Battle Mountain.

On July 4, 1891, Stratton staked out two full-lode claims on the granite formation that had had him restlessly viewing visions the night before. The two claims were named the Independence and the Washington. From the ore that he had sent to Professor Lamb of the Colorado School of Mines in Golden, he received an assay stating that his samples were worth $380 a ton. Winfield Scott Stratton had just struck it rich. He had no solid perception of just how prosperous his mines would be, but he knew that his luck had taken a radical turn.

By the spring of 1892 Stratton was well on his way to becoming the district's first millionaire. Within a very short period he was taking out $120,000 a month from his Independence Mine. He did not act recklessly as

Horace Tabor, the King of Silver, had done when his Matchless Mine began producing millions in silver during the 1870s. Stratton treated his gold mine like a bank. From the point where the Independence Mine began paying in great dividends, he never took out more than $120,000 a month from it. He held to his belief that gold was worth more in the ground than out.

The equipment the miners used was often more sophisticated than the quickly built structures which housed it.

Stratton hired friends he could trust to help him with his mine and paid them exorbitant wages. All employees were sworn to secrecy. Rumors that had slipped out told of veins uncovered at relatively minor depths which had millions of dollars worth of ore showing at the surface. The farther the miners dug into the claim, the richer the veins became.

After Stratton had successfully fought the claim suits against him when the wealth of his mine was made public, his claim on Battle Mountain grew to 110 acres. The mine had two major shafts, and its underground tunneling grew to more than six and a half miles. The largest veins in the mine were named the Bobtail and the Emerson. In December 1898 it had been calculated that 41,694 tons of gold ore had been taken from the mine with the average assay registering at $92 per ton.

THE WOODS EMPIRE

Warren Woods raised his two sons, Frank M. and Harry E., in Morrison, Illinois. Together their various mercantile businesses had acquired them enough money to stake silver claims in Leadville, Colorado, in 1878 — the Idalia and Black Diamond mines. But their funds were depleted before they could show a profit. Harry Woods returned to Warren, Ohio, and engaged in real estate and newspaper businesses. Warren and Frank also returned to Ohio and set up a drug store.

In 1888, Harry Woods was back in Colorado forming a brokerage firm and investing in Creede silver mines.

Warren and Frank Woods returned to Colorado in 1893 on encouraging words from Harry. Frank Peck of the Portland Gold Mining Company had interested Harry in the Cripple Creek area. When Warren, Frank, and Harry were once again reunited, the Woods Investment Company was established.

They were not in the area very long before they bought J.R. McKinnie's Mount Rosa Placer for $1,000. This purchase of 160 acres of land was platted as a town named after an old homesteader of the area, Victor C. Adams. The Woods promoted their town of Victor by claiming that every lot was a gold mine. As it turned out, one of the lots was a gold mine — for them.

The Woods company's first promotion was the Black Diamond Mine. However, their financial backbone turned out to be the Gold Coin. Upon this six-acre claim within the Victor city limits, 190 miners were employed to work. The first find while excavating for the Victor Hotel had been only a surface pocket of ore. But the Woods continued to develop the claim on a twenty-four hour basis.

The machinery on the project was quite impressive for the year 1900. The gallows frame rose to a height of 65 feet from a solid 24-foot square masonry block. The shaft had three compartments with a double-deck cage working in each. The hoisting engine was a double-reel, 750-horsepower dynamo which lowered the three cages in the shaft to the maximum depth of 1000 feet. Other pieces of machinery included a 24-drill Rand compressor and four Babock-Wilcox water-tube boilers powered by two batteries of 550 horses each. The total cost of the machinery was around $150,000.

From the first level of the Gold Coin, electric trollies and ore cars ran on three tracks under 3,700 feet of mountain to the Economic Extraction Mill in Arequa Gulch. The veins of the Gold Coin averaged eight feet in width with an average value of $46 per ton. The mine was also one of the few in which the largest stopes were filled in with waste rock.

The Woods were noted for their honesty and their generosity — both attributes they kept through their entire careers in the district. They were also known for the top wages they paid their employees. Their gratitude for their successes was extended to the men that worked for them in the shape of an elaborate meeting hall called the Gold Coin Club, which cost the Woods Investment Company $40,000. All of the employees knew that if they needed some kind of assistance, the Woods would be happy to help.

On Diamond Avenue, in the middle of Victor, the ornate shaft house of the Gold Coin could be seen. The beautiful pressed brick and stained glass of the building impressed all that saw it. The Gold Coin's location made it convenient for the miners to live in Victor and be close to work as well as home. But, the Woods encountered a large problem with the mine being located in the middle of town. There were fewer and fewer places to dump the waste rock from their mine.

The Gold Coin Mine in the center of Victor.

They decided to bore a tunnel from the Gold Coin, under the town, under Squaw Mountain, and into Arequa Gulch. This cost the Woods $200,000. Then they decided to build a chlorination mill, in Arequa Gulch, to process their gold ore. This cost $532,000. As the Gold Coin expanded, the Woods thought it would be more efficient for them to have their own power source. So they built one of the first hydroelectric dams in the country at Skagway, 11 miles south of Victor. This plant powered the underground system of electric tram cars through the drifts of the Gold Coin and the Economic Extraction Mill in Arequa Gulch.

The profits that the Woods Brothers were turning were astounding. The Woods Investment Company soon employed more than 1,200 men. The

Woods began spending their profits and their credits on obtaining every mine, mill, and property they could possibly buy. Their plans for a consolidated empire surged.

They increased power supplies to Victor and Cripple Creek with their Pikes Peak Power Company, which supplied 10,000 horse-power to the district. They powered more mines, interurbans, and hoists than any establishment in the district. Their Golden Crescent Water and Light Company brought water and electricity to Gillett and the surrounding mining properties. Soon the townsite of Grassy was bought from the real estate firm of Bennett and Myers for $180,000. It became the town of Cameron.

The Woods had their empire. Mines such as the Wild Horse, Deadwood, Pinnacle, Damon and W.P.H., Bull Hill, and Straub Mountain group were all within their United Mines Company which stretched out for more than four miles. Their companies included the towns of Victor and Cameron, the Mount Rosa Mining Company, the Economic Gold Extraction Company, Pikes Peak Power Company, the First National Bank of Victor, Pinnacle Park, and the United Mines Transportation Company. The Woods overextended themselves and everyone knew it.

One of the first banks in the townsite. It didn't take long to build fortified brick structures with thick cement and steel safes.

They might have been able to stay on top had it not been for disastrous events. The fire of 1899 in Victor set the Woods back considerably. Then their game of power in Pueblo over electrical companies was trumped by the

Thatcher family. Cameron lots were not selling as expected. Rumors of business troubles caused a run on their bank in Victor. And their hydroelectric plant at Skagway was costing more to maintain than it was producing.

Then — worst of all — the rich veins in the Gold Coin Mine became sparse.

Their empire crashed in 1902. After building Victor, Cameron, Pinnacle Park, and promoting almost a hundred mines as well as employing more than a thousand miners, the Woods' empire perished.

Frank Woods, who had been the director of mining operations, was to endure more pain in his life than just the loss of his $45,000,000 business. He lived through the deaths of his two wives, two sons, and a daughter. He died a pauper in Los Angeles in 1932.

Harry Woods, the company's administrator of finance, regained some of his crumbled empire and fled to California. He died in Laguna Beach, California, in 1928.

THE DISCOVERY OF THE RICHEST MINE IN THE DISTRICT

Johnny Harnan joined Stratton's favored mining group as a cook. One day Harnan climbed Battle Mountain looking for his own piece of land to stake out. At the top of the mountain he met a plumber named James Burns and a carpenter named Jimmie Doyle. Burns and Doyle were just about played out with their Little Portland claim. But Harnan saw something in the tiny plot of earth. He asked for a third share in the claim if he could get it to pay. The three Irishmen went into a partnership and soon struck ore assaying at $640 a ton.

They had to remain silent about their strike. Their little triangular claim of one tenth of an acre was bound to be jumped by the surrounding larger ones unless they had the financial resources to secure their property from the inevitable lawsuits that would threaten them as soon as their rich strike was made public. The three Jims decided they would take ore out of the mine in total secrecy until they had enough money stashed away. A month passed while ore was carried out of the mine in sacks by the three men in the middle of the night and taken to mills in Colorado City and Pueblo. One night a wagon in which they were smuggling out their own gold ore broke down. The wagon was so heavy it could not be moved before morning. By the first light, their secret was out.

Upon hearing of the rich strike up the slope from him, Stratton acted quickly. He offered the Irishmen $75,000 to add to their accumulated

$70,000 reserve so that they would be able to fight the impending lawsuits. But Stratton had not acted out of total generosity. He was made one-third owner of the Little Portland claim.

After defeating or settling 37 court suits at a cost of more than $280,000, the three Jims and Winfield Stratton were in complete possession of the Portland and its newly expanded 260 acres. The financial and strategic move had been impeccable for Stratton. The Portland Mine was destined to be the largest producer in the entire mining district — a figure of more than $62,000,000 in gold.

The total amount of ore taken out of the mine by December 1899 was 148,139 tons. The Portland was one of the best equipped mines in the country. The cost of its machinery was well over $160,000. The mine had three shafts with three compartments in each. All three shafts were connected in the underground workings of more than 66,250 feet. With its thirty different veins, the average ton was worth $50 in gold.

In 1899, the dividends to the stockholders were 24 percent of the original par value of the stock.

MORE GOLD RUSH TALES

The Gold Rush of the Cripple Creek-Victor Mining District was in full force by 1891.

Not long after the Portland strike, a druggist from Colorado Springs, A.D. Jones, had come up for his share of the loot. Knowing absolutely nothing about mining, he merely tossed his hat into the air and began digging where it landed. He established the Pharmicist Mine and was a millionaire at the age of 39.

Van E. Rouse was a school teacher who had obtained a third share in the Elkton Mine after its original locator sold half of it to pay a $36.50 grocery bill. During his free summer months, Rouse found himself in the middle of the gold rush trying his luck. At age 36, Rouse became another of the Cripple Creek-Victor millionaires by striking a rich vein in the Elkton.

In the meantime, Crazy Bob Womack, who had plodded around the region since 1878, sold out his El Paso claim for $500. The El Paso became the Gold King Mine and paid out more than $3 million in ore to its new owners. Crazy Bob had been ridiculed for nearly 15 years because of his belief that the Cripple Creek area contained a storehouse of gold. Crazy Bob was often seen speculating throughout the area during its first gold rush peak, but he died in Colorado Springs on August 10, 1909, as broke as he had been in 1878. Only Winfield Scott Stratton had remembered the founder of the "World's Greatest Gold Camp" with any real kindness.

Stratton had bought into the First National Bank of Colorado Springs with some of his profits from his Independence Mine and kept an office in its building. On sunny afternoons, Stratton could usually be seen sitting outside the building by the horse trough in front of the bank, cleaning his custom-made boots. Crazy Bob happened to be trailing down the same street on which the bank had been built one day just before the turn of the century. Stratton saw him. He picked himself up and made a beeline for Crazy Bob. Greeting him warmly, Stratton asked Bob how his prospecting was turning out for him. Upon hearing that Bob was low on his luck, he took out his checkbook and wrote Bob a check for $5,000. Then, with few other words passing between the two men, Stratton went back to his bank and took his seat beside the horse trough.

Victor was growing because of the huge production of Battle Mountain. The Woods had been profiting greatly from their Gold Coin Mine and their other interests. Stratton, whose wealth was already incalculable, began buying up land in Victor, old Fremont, Anaconda, and other areas. In 1896, the towns of Victor and Cripple Creek were rivals. The competition was more than just for financial benefit. The entire world was hearing about the region, and each community of the district wanted to be the front-runner. Victor had the mines. But Cripple Creek had the financial centers.

At least 30 millionaires came out of the Cripple Creek-Victor Mining District, but not all of these derived their fortunes from the initial gold rush. Most of the first generation of prospectors had either been veterans of the silver crash and the resulting panic of 1893 or were curious tenderfeet from Denver and Colorado Springs. The next "prospectors" were entrepreneurs who made the district the most elegant gold camp in the world — a gentleman's gold camp.

These entrepreneurs gained healthy incomes through the consolidation of the needs of the mine-owners and the miners. Their businesses included everything from A.E. Carlton's coal delivery service to the 24-hour gambling saloons and parlour houses of Myers Avenue in Cripple Creek. The mining district in the mid-1890s was like none other in the world. It possessed efficient water systems, Edison's light bulbs, the busiest telephone line in the nation, stock exchanges, influential banks, electric tram cars, bookstores, dry goods stores, sewer systems, and a multitude of daily newspapers. The restaurants offered cuisines with such items as fresh lobsters, strawberries, and pastries. Within a few short years after Stratton's Independence discovery, Crazy Bob Womack's dream had turned into a fortune chase for some 35,000 people.

The year 1896 had seen what had been thought to be no more than a worthless cow pasture turned into inexhaustible gold venture producing more than $500,000 worth of ore a month. And the first pinnacle had not been reached.

VICTOR VERSUS CRIPPLE CREEK

Besides Stratton's Independence Mine and the business dealings of the Woods Brothers, Victor was a convenient location for a meeting point of the railroad lines. Because of the many railroads into and around the district, Victor and Cripple Creek were unique as gold settlements of the Nineteenth Century. The wealth of Battle Mountain was bringing the ultimate in Twentieth Century science to the Old West. And because of this modern technology brought in by wealth, the Cripple Creek-Victor Mining District was never a "cowboy gold camp," but rather a civilized sort of high-society encampment.

This collision of two centuries caused uproarious times in the communities of Cripple Creek and Victor. The Old West lived on in the first generation of prospectors' hearts, but the second generation belonged to the incoming century.

With the arrival of the second generation of "prospectors," life became luxurious in the district, sporting such hospitalities as Turkish baths in the National Hotel of Cripple Creek and the baptizing of the new Cripple Creek Stock Exchange in foreign wine. Cripple Creek was vying for acceptance by the elite of Little London. However, the town of Victor continued to possess a heart of the Wild West within its city limits while tasting the fruits of the new era.

The town of Victor had the heart of gold and the scenery of the old West.

This difference in the life styles of the two communities tended to give Cripple Creek a favorable name while the reputation of Victor slipped backward into history. Both of these major communities of the district were lively places. But each town had its own standards.

Only five years back, in 1891, Pete Hettig opened the first saloon in Cripple Creek. It had been a tent which covered two bottles of whiskey, two barrels of beer, and a meager supply of cigars. Business had so flourished that Pete's friend, Bob Work, moved in a chair and established the first barbership. Around the same time, E.C. Gard published the first newspaper, *The Cripple Creek Crusher,* in gilt print.

Now in 1896, a half million dollars in gold a month was being shipped out of the area. Mining companies totaled 1,250 within the six square miles of land behind Pike's Peak.

The citizens of Cripple Creek wanted to form a new county and have their town be the county seat. In 1894 Cripple Creek and Victor had been of similar dimension and size. In 1896, the citizens of Cripple Creek invited Colorado legislators up to visit the town and consider the possibility of making it the county seat. The train, which had three specially chartered cars with the legislators inside, had to stop in Victor first to unload cargo. The citizenry of Victor, having heard that the legislators were aboard the train and their purpose for coming, detached the special cars from the train. When the railway engine moved on to Cripple Creek, the legislators were

The Twentieth Century "metropolis" of Cripple Creek. Population: 35,000.

being shown pleasantly about the City of Mines. The train arrived in Cripple Creek greeted by great fanfare, but the town's large welcoming committee found the legislators missing.

The battle was on. Despite the Victor citizenry's noble attempt to persuade the legislators, Cripple Creek became the county seat of Teller County in 1899.

There was not an area of life or business that was not included in the war for supremacy. Victor had Battle Mountain — nearly all that it needed. Cripple Creek could not contend with her sister town in that respect. But Cripple Creek "acquired" these riches of gold by becoming the financial center for the entire district. Bennett Avenue, in Cripple Creek, had amongst its assets three mining exchanges, countless real estate agencies, two major banks, the services of seventy lawyers, branches of all the major stores in the state such as the May Company, and many other businesses. Bennett Avenue, as a business sector, stretched from Poverty Gulch at its easternmost boundary almost up to Iron Mountain to the west. Cripple Creek's Bennett Avenue, by 1896, was definitely the district's major business section. So, in essence, money was mined out of Battle Mountain over in Victor then shipped to Cripple Creek to be disseminated to Colorado and the rest of the country.

Victor's business district was almost as large as Cripple Creek's at one time, but the town was never able to wield the influence that Cripple Creek had been able to. The publicity and contacts of Bennett and Myers, De La Vergne and Palmer, and Penrose and Carlton no doubt had engineered as much glamour and refinement as they possibly could into the town in which they had the majority of their business interests. Although Victor's business district was not as popular and prosperous as that of Cripple Creek, Victor's streets were literally lined with gold. No matter what kind of publicity came out of Cripple Creek, it could never claim, as did Victor, that it was the City of Mines.

Many of the roads in and around Victor were actually paved with gold ore. This low-grade ore — when the cost of milling rose to unprofitable levels — taken out of most of the mines was found to fetch a decent price from the highway department. Thus, many of the newly graded streets and roads in Victor became extremely rich to walk and ride upon. Some roads to the Portland Mine were assayed at a worth of $12,000 a mile.

The rivalry continued between the two largest towns in the district. Nothing was left unaffected by this competition for power. Even the tenderloin sections, especially Cripple Creek's Myers Avenue, fought each other for the greater portions of the entertainment business.

In their humble beginnings, one room "cribs" were prominent throughout the district. Every town had them. These cribs were the homes and businesses of the "fallen angels" of the area. Many of these women discovered their own "gold mines" in the business of prostitution. Many women found only dissolution, addictive laudanum, opium dens, and finally suicide through overdoses of morphine after engaging in prostitution. But there were few women in the district — except for the wives of miners — in comparison to the number of bachelors arriving daily. As the camps of the district grew into towns and the competition between Cripple Creek and Victor increased, these cribs were often combined to form larger parlour or sporting houses. These were directed by such famous madames as Lola Livingston and Pearl De Vere.

One of the most famous of these brothels was the Old Homestead in Cripple Creek. In order to be a profitable business, the Old Homestead had to elaborate enough to attract wealthy customers. As it did so it gained the reputation of serving the richest and most respected mining men in the region. Its decor had been nothing less than elegant, displaying rooms of French and English designs. Women were drawn from all over the world. At 353 Myers Avenue, it had become known throughout the world that the most expensive and refined pleasures were to be had by anyone who could afford them.

While the tenderloin districts had supposedly made it safe for the respectable women of the town to walk to bakeries and dance schools, the gambling saloons provided places for miners to cast their fortunes away as soon as they had dug them from the ground. Both Victor and Cripple Creek had saloons that remained open day and night. The district had more than 150 saloons and gambling houses. The names ranged from the humorous to the obscure: Bucket of Blood, Crapper Jack's, Last Chance, Manitou, Red Onion, Topic Theater, Smith and Peter's, Cabinet, Green Light, Midway, Silver Slipper, Turf Club, Red Light Saloon, Gavin and Toohey's, Abbey, and a hundred others. These "palaces" offered roulette wheels, dancing, liquor, opium dens, and women. The high life was to be had in every town in the district no matter how large or small, reserved or open, a miner or mineowner did not have to look far to find his pleasure.

Gillett — at the entrance to the district — was said to have been a bachelor's town, having the most concentrated entertainment sector in all of the southwest side of Pike's Peak. It never housed more than 800 people, but the town of Gillett had 25 saloons, 5 dance halls, and many sporting houses. Platted January 8, 1894, in Beaver Park Valley above Cripple Creek, Gillett had been the best placed townsite in the entire region. Advantageous water resources, train stops, and fertile valleys made excellent prospects for the town. Rich ore had even been found in the area. Unfortunately, this Lincoln strike was tied up in a litigation of claims for so long that the claim was never developed.

Much of Cripple Creek's water supply came from the Gillett area, sent down in wooden, metal-banded pipes. Gillett had names other than "Bachelor Town." It was also called "Gateway City" because it looked down on Poverty Gulch where Cripple Creek started. It was also called "City of Destiny" because it was the first town one came upon when entering the district on the Midland Terminal Railroad.

Gillett was a wide open town, housing more gambling and other "night time" pleasures per capita than any other town in the district. Today, only the Chapel of St. Dimas remains to mark its place in history.

Gillett had a half-mile race track where thoroughbred horses and dogs were run. Gillett's other unique feature was its bullfight held on August 24, 1895. It was supposed to have been the only known and publicized bullfight to take place in America. Promoted by Joseph Wolfe — a man always one step ahead of his creditors — the scheduled fight ran into problems from its inception. Francis Hill, the local secretary of the Colorado Humane Society, upon hearing of the scheduled event, protested the bullfight to the U.S. Secretary of the Treasury. Numerous warnings and threats were launched against Wolfe. These threats only served to spur Wolfe and his Wild West friend, Charlie Meadows, on. Carpenters went to work on the Gillett race track to convert it into a bull ring, and Soapy Smith — the infamous Denver and Creede gambler and con artist — bought the concession rights to the event. Tickets were offered at $2, $3, and $5.

Wolfe had the unsung support of many of the Cheyenne Mountain Country Club members such as Count Pourtales, so he felt safe in proceeding with his show. The finest bullfighters from Mexico were hired, including matador Jose Marrero. Reports stated that 3,000 to an optimistic estimate of 50,000 people showed up for the bullfight. But all of the bulls bought in Mexico had been stopped at the border. Wolfe had to improvise. Local bulls were brought into the ring instead. Three of these were killed, and the rest refused to fight.

On August 28, 1895, the entire Wolfe entourage spent a gala day in the Colorado Springs jail, guests of Francis Hill and the U.S. Secretary of the Treasury.

THE END IS STILL NOWHERE IN SIGHT

The exotic and extravagant nature of the Cripple Creek-Victor Mining District increased as the production of gold ore soared to new peaks. Electricity replaced all of the old power sources before the turn of the century. It was the only mining district in the world that could boast that its miners rode electric street cars to the mines, were lowered down the shafts by electric hoists, their way lit by electric lights, while drills were electric air-powered and dynamite was set off at electric control panels connected to the charges by cable.

The electric trolley service had a "High Line" and a "Low Line" which connected every town, camp, and major mine together. Fifty-six trains ran daily on the surface, and many more ran ore underground to mills, shaft houses, dumps, and chutes. Coal had once been the main source of energy. A.E. Carlton started his Colorado Trading and Transfer Company with his brother, Leslie, in 1893 by bringing coal in wagonloads up to each mine. By the year 1898, the Colorado Electric Power Company of Canon City was generating power up 24-mile transmission lines to a central distribution station in Victor.

Power sources were not the only radical changes in the region. Ore reduction and processing was improving. Chlorination leaching mills had sprung up all over, but a more efficient method of extracting the gold from the granite and breccia was needed to keep production costs down. The Economic Gold Extraction Company Mill was built by the Woods Brothers in Squaw Gulch. This was the largest reduction plant of its kind in the United States. The Columbine-Victor tunnel was bored in 1899 from the Gold Coin Mine, under Squaw Mountain, and directly into the plant to save production costs. But the mill used a chlorination process which was rapidly becoming too costly to allow the mining to be profitable. Experimental plants using bromide and cyanide were just starting to get their first major boost around the turn of the century.

The cyanide method, modified from a process used in Scotland since the mid-Nineteenth Century, proved most efficient, making profitable even the processing of ore which assayed at $10 a ton. The speculative worth of the entire district increased again.

As soon as the problem of processing had been temporarily solved, a new dilemma arose. This problem was with the groundwater tables of the district. The deeper mine shafts were sunk, the more susceptible they were to flooding. The first of many tunnels bored to drain off some of these subterranean flows was named the Ophelia Tunnel. This tunnel opened the base of Gold Hill and was dug like a drift for 2,600 feet. Water discharged out of its mouth at 2,000 gallons a minute, and the hill went dry in 1898. The second of these tunnels was the Standard Tunnel. Upon its completion, 2,800 feet into the ground, a flow of 1,800 gallons per minute was tapped. Between the Ophelia and the Standard, both Gold and Raven hills were completely drained by 1899. These two tunnels set the formula for the major tunnels to come, such as the Roosevelt and the Carlton, which in their prime would give the district many decades of extended life.

There were nearly two hundred paying mines in the district registered in the years between 1896 and 1899. Teller County was formed around the mining district, totaling 555 square miles. The Woods were nearly finished with their hydroelectric plant at Skagway. The Smith-Moffat company had an electrical generating plant at Goldfield. Gold production had soared to $18 million at $20.67 an ounce for the year of 1899. And the population of the entire district had surpassed 35,000.

Just outside of Gillett, a lone Italian priest who could barely speak a word of English had finished the single-handed construction of his Saint Dimas of the Mountains Chapel. Today, its stone foundation is nearly the only structure left to indicate the presence of the town of Gillett.

THE SOCIALITES

Mysterious and charismatic characters were constantly being attracted by the opulence of the Cripple Creek-Victor Mining District. Such bunco men as Joe Wolfe and "Parson" Tom Uzzel, such business moguls as D.H. Moffat, and unpredictable sorts such as Winfield Scott Stratton came into the region and left their marks. A certain group of men came in with the second rush of "prospectors" who were generally from upper-class families of the East. All of them had excellent educations. They also maintained powerful wills to succeed in whatever they undertook. They were known as the Socialites.

The Socialites were a hearty lot, all were as interested in making careers for themselves as they were concerned about enjoying every minute of life that they could. Most of them started humbly in Cripple Creek and Victor. Not one of them was near the age of thirty. And every one of them struck it rich, one way or another.

The most prominent of the Socialites was Spencer Penrose, brother of Dr. R.A.F. Penrose. Spencer Penrose was a man who could not be satisfied with pursuing only one particular interest. In 1893, he was part owner of the C.O.D. Mine. In that same year he was managing the properties of the firm of Tutt and Penrose. Also during this time he was formulating ideas for new processing mills for the tellurides found in the district.

Spencer Penrose's rich career began in Cripple Creek. His crowning achievement was the Broadmoor Hotel.

The Socialites were intelligent, social, handsome, and lively. Their mad antics in either the business sector during the day or in the "tenderloin" district at night were told and retold by all. But they were not all play and frivolity. Between Spencer Penrose, Verner Z. Reed, Charles MacNeill, Harry Leonard, Charlie Tutt, and Harry Blackmer — Socialites all — their total aggregate wealth would soon approach $100 million.

Sally Halthusen, an attractive blonde, had found her way into the Socialites. Sally was most noted for her beautiful features and the majestic white horse that she always rode about town. Unlike the other respectable ladies in the town of Cripple Creek, Sally refused to ride side saddle. She always caused a stir in the saloons as she rode by to visit one of the Socialites' offices. She loved horses and wanted to marry a man who could provide the wealth to buy her great animals and stables. All of the Socialites courted her and accepted her as one of their group. However, Sally's fortune was not to marry Penrose, or MacNeill, or Leonard. She married a confectioner who owned land in the region. Her great white horse, upon its death, was buried in a grave blasted out of the granite strata on Mt. Pisgah. The madame of the Old Homestead, Pearl De Vere was also buried in a grave in the cemetery near Mt. Pisgah. Madame De Vere's original wooden grave marker is now in the Cripple Creek Museum. A marble tombstone has replaced her vulnerable wooden memorial.

Spencer Penrose accumulated his interests in the district around the turn of the century. They included the C.O.D. Mine, milling investments in Canon City, and the race track in Gillett. Later, he placed these investments in a copper-milling plan in Utah. A metallurgist, Daniel C. Jackling, had a plan which would revolutionize the milling industry. A few years after this major investment, Penrose acquired for himself the income of $200,000 a month for life. But Penrose did not forget where he had made his first sizable chunk of money. With some of their profits from their copper mill in Utah, Spencer Penrose and Charlie MacNeill built the Broadmoor Hotel in Colorado Springs. And, with all of the influential power that Penrose had acquired, he helped his friend A.E. Carlton take major control over the Cripple Creek-Victor Mining District. But this is jumping ahead to the years between 1910 and 1916.

The influence of the Socialites was still in its infancy in 1899. But that would soon change. A.E. Carlton's Colorado Transfer and Trading Company had improved until he was able to sell it for a profit so that he could buy the First National Bank of Cripple Creek. From this advantageous pedestal, Carlton was able to uncover a myriad of interesting facts about many influential businesses and their property values. He soon began buying up mines and other concerns which would lead to the building up of his empire — one that would ultimately be worth $230 million in production values.

The year 1906 saw the untimely flooding of the El Paso Mine. The stock value of the mine dropped more than a half million dollars. Other mine owners became worried that their properties would be the next to be devaluated. The El Paso Tunnel had been overburdened by the spring runoff of the heavy winter snow. A new tunnel had to be built. A.E. Carlton was in a position to propose a $500,000 tunnel which would be drilled through Gatch Park. The other mine owners protested that the granite was too hard in that area and that an earlier attempt at boring a tunnel had failed. Carlton was persistent. The other mine owners finally agreed with his proposal.

From 1907 to 1910, Albert E. Carlton planned and managed the drilling of the Roosevelt Deep Drainage Tunnel. When the tunnel was finished in 1910, its full length was 15,740 feet and drained subsurface water flows at 8,500 gallons per minute. The cost had been $815,000. But it had added years to the life of the mines in the area. A.E. Carlton was then considered one of the most influential men in the district.

The Socialites continued to amaze and delight the people of the region for many years. Their brilliant plans, wealthy strategem, and fast-paced lives were of those that made novels best sellers. Their standing in the communities of the area helped the reputation of Cripple Creek more than anything else.

When Verner Z. Reed sold Stratton's Independence to Stratton Independence Ltd. of London, the transaction placed Victor on the map beside

Cripple Creek. From this point on, no other mining district throughout America was so closely bound to English capital until the South African gold mines gained prominence.

Reed was born in Richland County, Ohio, in 1864. By the age of twenty-three he was already involved in real estate promotion. He brought the idea of constructing houses and selling them on installment plans in Colorado Springs. Reed had mining interests in the Cripple Creek-Victor Mining District, but it was not until he helped clear the Portland mining claim that he gained his firm reputation. From then on, he worked only for the richest and the best: Stratton, Burns, McKinnie, and Harnan.

Reed and C.C. Hamlin formed the Reed and Hamlin Investment Company, which had offices in Cripple Creek, Colorado Springs, Arbroath, Scotland, and on 58 Lombard Street in London. From his position in London, Reed encouraged the Venture Company to take up the option on Stratton's Independence mine in 1898.

Reed ended up selling the Independence for 666,666 shares of stock in the English company of Stratton Independence, Limited. But this transaction only took place after Colorado State geologist T.A. Rickard inspected the mine for five weeks. After his investigation, which included taking 1,008 ore samples weighing an average of 30 pounds each, he stated that the Independence Mine was still worth at least $8 million.

As the shares of the Stratton Independence, Limited, were traded on the stock exchange, Winfield Scott Stratton quickly received his asking price of $11,000,000 for the mine. From this, Reed gained $1,000,000 in commission. Verner Z. Reed then moved to Paris for several years, living the life of a rich Bohemian. Upon returning to America some years later, he bought a 700-acre ranch outside of Sheridan, Wyoming. There he invested in the Salt Creek oil fields and made his real fortune.

FIRES TILT THE COMPETITION

The mines of the Cripple Creek-Victor Mining District were always filling up with water, but the surface of the plateau was exceedingly dry. Expensive pumps and then expensive drainage tunnels were used to keep the mines free of water as they reached deeper and deeper levels. But less care was taken in preserving the towns.

April 16, 1892 brought a harbinger of things to come. One half of a city block on Bennett Avenue burned, destroying $10,000 in buildings and goods. This warning of 1892 was not heeded. Although this fire was minor compared with the fires that were to take place in Cripple Creek in 1896, little was done to prepare for future disasters. Water reserves on the surface of the district were low, and much of the construction in the towns was of dry timber.

In 1896, Otto Floto, bartender at the Butte Opera House, was arguing with his girlfriend in a room on the second floor of the Central Dance Hall in the tenderloin district of Cripple Creek. As the vocal argument grew to physical violence, a lit gasoline stove was sent crashing to the floor. Within minutes, the three revolver shots which signaled the fire alarm were heard throughout the town.

By the time the fire department had tapped what they could from the reservoir, the Topic Dance Hall next door to the Butte Opera House was also on fire. The light breeze and warmth of April 25, 1896, was not helpful to the firemen. More saloons, parlour houses, and cribs caught fire as if they were matchsticks. Shacks had to be dynamited in an area on Third Street between Myers and Bennett avenues to prevent the further spreading of the fire. Soon the flames were feeding upon the First National Bank and the Cripple Creek Mining Exchange. In three hours, one-twentieth of the town had been burned. The estimated loss of property and goods was more than $500,000.

The recurrent Cripple Creek fires of 1896 convinced many people to rebuild in the most expensive material of brick. After all, Cripple Creek was a "Gentleman's" gold camp.

Wednesday, April 29, was a day of cleanup and rebuilding. Cripple Creek was determined not to be out of business for long. In the afternoon, those fateful three shots were heard once more. The old Portland Hotel was on fire. The building was so fireprone that the roar of the flames could be heard as far away as Altman. New forceful winds blew in. More businesses and residents on Bennett and Myers avenues went up in flames. The firemen found the town's reservoir to be nothing more than mud. Then the boilers in the Palace Hotel expanded past the danger point and exploded. Five firemen were injured. Not long after this explosion, seventy pounds of dynamite in the Harder Grocery sent a surge of flame high into the air. The whole town shook with the force of the unified blast. And more saloons, shops, stables, stores, and homes burned. The Victor Fire Department had shown up to lend its men and equipment, but what was needed most was water. As the flames consumed great portions of the town, more than 5,000 people were left homeless.

News of the fire hit Colorado Springs. Stratton and the Cripple Creek Victor millionaires living on Wood Avenue quickly gathered their resources and sent aid to the area. All of Colorado Springs immediately followed suit. In a few days after the second fire of April 1896, most of the stores and saloons were back in business as they had started out in 1891 — under tents.

During this time of the Cripple Creek fires, Victor accepted its role as the leading community in the district. But this major role was short-lived. Cripple Creek did not take very long to get back upon solid foundations. Most establishments in Cripple were rebuilt in brick. Two months had barely passed when a large section of the town's businesses were back in the local economy.

It took nearly a year, but 107 new brick buildings were built and a wave of massive improvement had overtaken the town of Cripple Creek.

Cripple Creek rebuilt mostly out of its own resources. Although it was well-known that the Cripple Creek-Victor Mining District had broken the National Panic of 1893, little substance came in from the outside world to bebuild the town. Less than 25 percent of the buildings and businesses burned had been insured. Reinvestment and the discovery of new mines in the western sector of the district aided the reconstruction. Victor's temporary dominance in the district was cut short by late 1897. It never regained the brief moment it held while Cripple Creek was rebuilding, even though 80 percent of the district's wealth came from the southwest area in which it had been platted.

Victor's greatest prize during this period was its Fourth of July celebration in 1897. As Cripple Creek rebuilt, it rapidly became even more of an elitist's society, displaying its new brick buildings along Bennett and Myers avenues as a haughty woman would show off a new dress and carriage. The residents of Cripple Creek did not think it proper to have a boisterous party on Sunday, even if it was the Fourth of July. The town of Victor thought Independence Day was a great deal more important than moral facades. People from all over the district crowded into Victor and enjoyed one of the richest and most elaborate celebrations the Fourth of July had ever seen.

All kinds of competition events were staged. Drilling contests for the hard-rock miners. Dance contests for the ladies. Shooting galleries. Ore-hauling sports. And a multitude of other festivities. Thousands of people crowded the streets of Victor. Foppery and proper society be damned. It was the Fourth of July in America, and the people of Victor had more than one kind of Independence to celebrate.

But all the guffawing at Cripple Creek appearances would not last long for Victor.

Victor was the shipping center for the gold ore of the entire district. Its riches were evident throughout the town. Men mined in their backyards, and wagonloads of ore rolled constantly down streets and alleys to the mills. In the year 1897, Victor looked as if it would gain the leading position for the district.

Just as the battle for the formation of a new county around the mining district and the new county's seat of authority was boiling into a fury, the town of Victor suffered its own catastrophic fire. Near Jenny Thompson's "999" dance hall at 307 Portland Avenue, a fire broke out in a poorly built wooden shack. The rest of the tenderloin block caught fire. Water levels were low in the reservoir, as usual. When the fire hoses were brought in to fight the fire, they ignited in the intense heat from the quick-burning wood shacks. The fire departments of Cripple Creek, Goldfield, Anaconda, and Altman responded to this fire on August 21, 1899. However, since water was scarce there was little the firemen could do.

Fire breaks were devised with dynamite. The fire and the dynamite ended up destroying 12 city blocks. Another 25 blocks were severely damaged. Yet, by the 23rd of August, businesses in Victor were flourishing once more. The town of Victor was rebuilt quicker than Cripple Creek, owing much of this expediency to the consolidated efforts of the Woods.

Everything about the recovery of Victor was brisk. A "fireproof" city was planned by the city council. Steel-reinforced, brick, and stone buildings replaced the wood shacks. Improvement was rapid. Even the tenderloin district moved from the alleys of Portland Avenue to First Street.

A $30,000 opera house which seated 1,200 people was built. Ironically, W.C. LeMaster, set fire to this brick building on September 27, 1920. LeMaster was the owner of the Victor Opera House at the time and was suffering great losses as the wealth in the district slipped into a decline during the third decade of the Twentieth Century.

The town also had a new city hall and a comfortable jail built of brick. Both of these were to become quite useful as more and more ruffians and rascallions were caught using the mines of Victor as their bases of operation. The Gold Coin Mine's shaft house burned and was rebuilt in ornate pressed brick. The reading room that the Woods Brothers had provided for the miners in the original shaft room was also rebuilt and new books and magazines stocked on its shelves. Some of the miners complained about the stained glass windows put in the shaft house. They felt as if they were in church when they entered into the mine and found it almost impossible to swear. A palatial club for the Gold Coin employees was constructed on Granite Avenue. Schools, theaters, and a new fire house were also on the list of reconstruction. But foremost was the improvement and expansion of the town's water supply.

The lobby of the Victor Hotel was a gathering place for all sorts: miners and millionaires, con men and mine owners, union men and speculators. Even the cultured English investors saw such gatherings as amusing entertainment.

However, the worst luck that had come from the 1899 Victor fire was the damage to the Woods' empire. They had overextended their estimated $45 million worth in their attempts to consolidate their mining operations. The fire and reconstruction cost them a great deal and pushed their empire to the edge of collapse.

Upon the death of Winfield Scott Stratton and the fall of the Woods' empire, the entire district suffered a great decline in production. Also, the rivalry between Victor and Cripple Creek ended. Both Victor and Cripple Creek had suffered great losses.

The Gold Coin Mine burns in the Victor Fire of 1899. The end was nearing for the Woods $45 million empire.

SECESSION FROM THE NATION AND THE BATTLE OF BULL HILL

Violence was unavoidable. Events and personalities all contributed to one of the most incredible strikes in American labor history. And the Cripple Creek-Victor Mining District was just the place to stage this sensational foray between worker and business owner. All the world was watching this gentleman's gold camp.

The gold of the area helped the country recover from the Panic of 1893. But the issue of gold standard vesus silver standard was still extremely volatile. The political philosophy of Populism and one of its major goals, the free coinage of silver, had swept the state of Colorado. Even a Populist politician, Davis H. ("Bloody Bridles") Waite, had been elected governor.

Most significant of all, the mine owners, realizing that their average pay for hard-rock miners was higher than any in the West, decided it would best fit their production costs by implementing a $3 per 10-hour work day.

The Western Federation of Miners (WFM) had been born in Butte, Montana, on May 15, 1893. Its infancy was already long behind its forward movement. John Calderwood, a man who quoted the words of Pope Leo XIII, organized the men of Battle Mountain above Victor into the Free Coinage Union, Local Number 19, of the WFM to combat the mine owners' new $3 per 10-hour work day. Within eight weeks, Calderwood had nearly eight hundred men in his union. Most of these were centered around the new town of Altman.

Altman was nearly 11,000 feet above sea level. In November 1893, Altman had a population of 1,200. There were the same number of saloons in the town as there were grocery stores. It had many restaurants and boardinghouses, a telephone service, and a school. Located near the rich mines — Victor, Buena Vista, and Pharmicist — Altman was in a prime position to become a haven for the disgruntled miners. The town looked down over the entire six square miles of Bull Hill and Battle Mountain. Its scenery was incomparable. Its strategic location was impregnable. Between storms and continuous dynamite blasting in the area, Altman seemed to be always rumbling with one sort of fury or another.

Through Calderwood, the miners protested the new work schedule put into effect by the mine owners. Such men as Dave Moffat and Edward De LaVergne thought Calderwood a laughable figure, not worth recognizing as existing. Winfield Stratton and Jimmie Burns were not so quick to underestimate Calderwood's power. When most of the mine owners signed an agreement to go to $3 per 10-hour work day — except for those such as Stratton and Burns — Calderwood called a general strike.

February 1, 1894, five hundred miners refused to report for work. Most of the Bull Hill and Battle Mountain area had to close down. Many of the mine-owners, who were residing in the El Paso Club in Colorado Springs, still laughed, believing the miners would only hold out for a few days at the most. But those miners that gathered around Calderwood in Smith and Peter's saloon in Altman knew different.

Calderwood was a leader and an organizer. He had gathered nearly $8,000 in funds for the striking miners to live on while not working. Then, after setting up a bread line and a central food-supply center, he taught the strikers how to anger the opposing side and still stay within the limits of the law. Yet, throughout, Calderwood remained in close contact with Stratton and Burns.

All of this was merely a prelude to a strike that was to last for four months, costing $3,000,000 in losses, and initiating the use of many tactics for future labor strikes.

Surprisingly, Stratton signed a labor contract with Calderwood. This set the El Paso Club into an uproar. The terms were for a $3.25 per 9-hour

A cramped and busy office of a Cripple Creek Hotel in 1901.

work day. At Calderwood's Wilson Creek cabin, Stratton thought he was doing what was best for mine owners and miners alike. The repercussions were not what he had expected.

The fractured yet still abundantly wealthy El Paso Club had Judge Campbell of the Springs issue warrants for Calderwood and one hundred others. They would have had Stratton arrested, too, had he not owned the Independence Mine, part of the Portland, and a multitude of other properties.

Sheriff Frank Bowers had to serve the warrants for arrest in Altman. He did not like his task, especially since he had won election to the office of sheriff through the help of the miners. On March 15, 1894, the warrants were served. The miners' reaction was to threaten to blow up the Victor Mine. Bowers had to send six deputies to protect the Victor. All six deputies were captured by the strikers.

Calderwood was serious about his strike. He had the deputies released, but when scabs were brought in to work in Moffat's and De LaVergne's mines, the town of Altman seceded from the United States of America. It was in an excellent position to do so. Its location at the very top of Bull Hill made it inaccessible when fortified. All around were the Fourteeners of the Colorado Rockies, with Pike's Peak being the closest. Calderwood formed the "Kingdom of Bull Hill" and issued passports and visas that were only valid with his signature. His stand was impregnable.

Sheriff Bowers called Governor Waite to send the Colorado Militia into the district. They arrived on March 18 and camped at George Carr's Broken Box Ranch. By this time, the miners at Altman had prepared for war. They had built a stockade to help fortify Bull Hill and constructed a huge bow which could catapult a beer bottle packed with dynamite up to a quarter of a mile. A fake cannon was pointed at the Victor Mine to frighten the mine owners into thinking they could blow up anything they wanted to. All the roads around Altman were guarded. Travelers were checked for passports and visas signed by Calderwood. Many of the roads were mined with fierce explosive charges.

While Sheriff Bowers was in Colorado Springs obtaining more warrants, General Tarsney of the Colorado Militia sent for Calderwood to meet him at the Palace Hotel in Cripple Creek. At the hotel, Calderwood and Tarsney made a deal. When Bowers returned with the warrants, he wanted Tarsney and the militia to accompany him up Bull Hill. Tarsney told Bowers that he would have no trouble and that he should go up to Altman alone to serve his warrants. Bowers had no idea that Calderwood was dealing with Governor Waite through Tarsney. Bowers knew his duty as sheriff came before his fears. He wrote a letter to his wife telling her that he might not return from Altman, then belted on his guns for the journey up to the Kingdom of Bull Hill.

Every step of the way he expected death. He had heard of the pencil-sized dynamite charges that the miners had been equipped with to kill their foes at close range. He knew that many of the miners were crack shots with either pistols or rifles. At the small town of Midway, everything appeared normal. That made him even more expectant of death. Warrants in hand, he proceeded up to Altman.

It was unusually quiet on top of Bull Hill. As Bowers entered Altman he was greeted by a cheerful Calderwood. Bowers was still prepared for an attack. Calderwood took Bowers into Smith and Peter's saloon and served him some of the finest bourbon to be had in the town. When Bowers brought up the subject of the warrants, Calderwood told him that all who were named would submit without a struggle to the arrests. Then Bowers knew he had been set up by Governor Waite, Tarsney, and Calderwood to make Judge Campbell and all in Little London look like fools.

The war appeared over. In April 1894, Calderwood went on a tour of the other Colorado mining camps for the WFM. During his absence, the relative peacefulness of the strike quickly ended. Calderwood had left Junius J. Johnson in charge of the "Principality of Bull Hill." Johnson did not have the control over the miners that Calderwood had. A criminal element began infiltrating the ranks of the strikers.

"General" Jack Smith became a prominent leader in Altman. His rage against the mine owners was dastardly. Once, after capturing two spies for the mine owners amidst the strikers, Smith ordered the two to drink from cuspidors. He then threatened them with mutilation. One was ridden out of

The Elkton Mine during its prime years. This mine and others surrounding it were so productive the town of Elkton was platted to settle the miners close to their work.

Altman disgracefully, while the other was thrown down a prospecting hole. Smith was a terror. He tore up saloons, assaulted nonunion men, disrupted the peace at parlour houses, and stole from any store convenient to him. His gang was infuriating the mine owners once more and making the whole district look ugly through the sensational coverage the newspapers gave him.

The mine owners at the El Paso Club decided on a plan that would show a distinct spirit of retaliation. Ex-policemen and ex-firemen from Denver were hired as a posse to go up to Altman and frighten the miners. "General" Johnson received information of the posse's arrival and arranged a reception for them. Coming up on the narrow-gauge Florence and Cripple Creek Railway — just as they approached Victor — the Strong Mine erupted in a powerful explosion. Sam McDonald, Charles Robinson, and some others were caught in the mine. Another eruption caused the steam boiler of the Strong Mine to blast into the air and then shower down in fragments upon the posse on the flatbed cars. The train stopped, shifted into reverse, and traversed quickly back down the mountain.

The mob of miners sent a flatbed car full of dynamite after the posse. Then, while still in a mass rage, they began looting stores in Victor, beating nonunion men, and destroying private property. Stratton's Independence Mine was captured, and the war was on again.

Later that night, an attack by Smith on the posse cost the lives of two men. Five strikers were captured by the posse.

Stratton initiated a new movement against the strikers because his "golden bank" had been taken. He organized and helped finance 1,200 deputies into an army that could take Bull Hill. William S. Slocum set up an exchange of prisoners. This took place on May 28, 1894, under the international laws of war. Sam McDonald, Charles Robinson, and the others who had been trapped inside of the Strong Mine were exchanged for the five captured strikers. Charles Robinson had gone permanently insane while trapped in the Strong.

Company "M" of the army of deputies Stratton had helped finance after his "golden" bank was captured by the striking miners.

Stratton was exploding with anger. He wanted his property returned to him. He had tried to help the miners before. He had been on their side in the earlier skirmishes. He had even come out in favor of the silver standard. Now he would never again trust the miners.

Winfield Scott Boynton, a Manitou Springs shoe clerk, led the 1,200 deputies up to Altman. Governor Waite stepped in for the miners, going up to Altman himself to be their spokesman. General Tarsney had been sent out to come between the strikers and the deputies. Three armies were maneuvering around Bull Hill. All three armies were costing the mine-owners a large sum of money to maintain.

The 1,200 deputies charged Bull Hill on June 6, 1894. Confident that they would win, they made a frontal approach. They were met with bullets and dynamite. Boynton had to retreat.

Tarsney placed his militia in Grassy Gulch, acting as a mediator between the strikers and the deputies. This battle position did not work well for Tarsney. The deputies attacked again on Friday, June 8. This time the deputies had to retreat down Globe Hill. Tarsney not only interceded after this retreat by Boynton and his deputies, but he succeeded in occupying Altman. The miners had asked him in.

Lined with contradictions, a peace ensued. On June 10, 1894, a treaty was signed. Calderwood stood trial along with 300 other men. All were acquitted except for the two who blew up the Strong Mine. The miners got a $3 per 8-hour work day as part of their settlement. The mine owners were allowed to hire whomever they wanted to work in the mines. After the signing of the peace treaty, a victory celebration ensued for days.

Not long after, on June 22, 1894, General Tarsney was drawn out of the Antlers Hotel in Colorado Springs by a hoax. The men who had lured him out of his rooms captured him and took him off to be tarred and feathered.

The McKinney Mine above Anaconda blew its top one day, and showered thousands of rocks over the town.

Various reports state that John Calderwood, after he was acquitted, either became an assayer or a justice of the peace in Victor, or disappeared from the area altogether. "General" J.J. Johnson became a colonel and died in the Spanish-American War. As for "General" Jack Smith, he escaped to Missouri for a time. Unfortunately for Victor, it was not the last that the district was to see of him.

THE CORE OF THE DISTRICT

The towns of Cripple Creek and Victor were still at odds for supremacy of the region. But Victor's luck began to turn sour. The rich mines that had made Victor "the core of the rich Cripple Creek Mining District" had become infested with vermin.

Mines such as the Gold Coin and the Strong had drifts and shafts that permeated the foundations of Victor. In all directions, and in a multitude of layers, mining tunnels had been bored under and around the town's city limits. These tunnels became havens for gangs. One such gang was headed by the notorious Jack Smith, along with his brother George and Dynamite Shorty McLain, the man responsible for blowing up the Strong Mine in 1894. Smith still enjoyed robbing stores and disrupting the peace of parlour houses, but his best game was high-grading.

Not only did the Smith gang use such mines as the Strong, Dillon, and Granite as hiding places from the law, they robbed the mines of their rich ore. It was easy for a man to live comfortably off of what he could high-grade from the mines. His greatest problem was knowing his way through the numerous shafts and drifts. If lost in the mines while high-grading, he could expect little help from anything human.

On a riotous spree one day, Smith decided to attack the town of Altman. Single-handedly, he captured the town and freed all of the prisoners in the jail. He retreated with his spoils to Victor. Back in Altman, Marshal Kelly made out a warrant for Smith's arrest. Kelly captured Smith in Victor.

No sooner was Smith in jail than he was out on bail. Word rapidly spread that Smith was waiting for Kelly at Gavin and Toohey's saloon in Altman. Kelly was not a man to avoid a challenge. He met Smith at Gavin and Toohey's saloon. A gun battle took place. Smith's aim was bad, and he ended up killing George Pabst, a member of his own gang. The marshal was more accurate. Kelly put a bullet through Smith just below his heart.

This incident did not end the crime under the streets of Victor. The violence was only beginning.

A woman named Hook-and-Ladder Kate took over the reins of the Smith gang. A great many of their schemes went askew. Robberies of Wells-Fargo wagons and the Bi-Metallic Bank were thrown awry by arrests and blunders. Then George Smith and two others were killed in a fight at the Union Theater. Some months later, Kate used morphine to commit suicide.

Henry McQuaan took control of the Smith gang. One night he decided to beat up the bartender at the Green Light Saloon. Policeman Frank Lupton arrested McQuaan after beating him senseless. A friend of McQuaan's, Seth Ralston, tried breaking McQuaan free from the policeman. Lupton fired one bullet, causing superficial wounds in both McQuaan and Ralston. All three were handcuffed together. A crowd blocked the entrance to the Victor jail. So Lupton decided to drag McQuaan and Ralston to the nearest train going out of Victor. The mob continued to haunt Lupton. Before the engine could depart for Cripple Creek, the mob tried detaching the car containing Lupton, McQuaan, and Ralston. Lupton fired out of the window and the mob dispersed. But they did not give up. Around Squaw Mountain at Elkton another mob boarded the train. Lupton protected himself with the bodies of McQuaan and Ralston and fired into the mob. With only one death, Lupton finally got McQuaan and Ralston to the Cripple Creek jail.

A typical saloon of the district. The more whiskey dispensed in coffin flasks, the more chewing tobacco ended up on the floor rather than in the spittoons.

The Smith gang was not the only underground force in Victor. The Crumleys were another. However, their luck in breaking the law was a little more prosperous. The Crumleys started out with saloons and gambling, moved onto train robbery, then found it safer to be bunco men like their "fence" Soapy Smith. One of the main attractions in their saloons was Grace Carlyle, who adored taking off her clothes atop bar counters. But the Crumleys wanted to move into something with a bigger future. They took up claim and ore swindling, and then went into high-grading, using the Strong Mine as their main holdout.

High-grading, the stealing of ore from mines, increased in popularity among thieves when a judge ruled that the taking of ore was not a crime because "real estate could not be stolen." A High-Graders Association was formed which included some thirty assay offices in which the ore could be unloaded. Many stole rich ore from the mines in which they worked. Miners took out ore in their lunch boxes, under their armpits, in the hollowed out handles of picks; by capturing drill dust in their bear-greased hair, or literally rolling themselves and their damp clothes in the fine gold talc. The ingenious methods devised were innumerable. A miner could make up to a $100 a day by high-grading when his usual 8-hour work day brought him only $3.

High-grading became a $5,000-a-day business. It was estimated that the Independence Mine alone lost $1,000,000 to high-graders. Soon mine owners installed "change rooms" where the miners had to strip and be searched before leaving the shaft house of a mine. Then high-graders had to become even more ingenious.

Miners would return to the shafts and drifts at night to steal ore. In a stope of the Independence known as the Jewelry Shop, a group of high-graders was surprised by the guards of the mine. Lights and candles suddenly went out and bullets rang through the drifts. Totally in the dark, one man

An assay office in Cripple Creek. Many of the instruments used in this process are on display in the Cripple Creek Museum.

was shot and another fell down a winze shaft. But, battles such as this one did not stop the high-graders.

The assayers involved in high-grading grew more and more greedy every day. They started taking more than half of the cut from the money on the sales of the gold ore. The miners could not complain because they had no one they could go to. So, when assay offices implicated in high-grading throughout the Cripple Creek-Victor Mining District began blowing up, it was difficult to pinpoint who the fuse-lighters were: the miners or the mine-owners. High-grading continued no matter what precautions the mine-owners devised, and the thievery and violence of the district literally went underground.

The only development that finally put a halt to wholesale high-grading was the scarcity of rich ore in the mines. But, on rare occasions, it was still possible to see high-grade ore in the collection plate on Sundays.

Crime, however, was otherwise very low in urban areas. Few people locked the doors to their houses. Even fewer could find the keys to their front doors because they used them so rarely. The town of Cripple Creek did not even have a jail until it became the seat of Teller County in 1899. Most of the illegal activities took place in and around the mines. Swindling, claim jumping, salting, high-grading — every con game was centered around gold and gold mines. The meager properties of the miners were generally

The Arno Mine within the city limits of Anaconda was one of the strongholds for the criminal element of the district. Otherwise, the urban areas were very safe for their occupants.

left untouched. The only times rampant crime occurred in the towns was when a gang held out in a mine that was within the city limits — such as the Strong Mine of Victor or the Arno of Anaconda. Otherwise, most illegalities were either misdemeanors or somehow related to labor warfare.

1900 - 1901

Considered the pinnacle of gold production for this era, 1901 saw the brilliance of more than just refined gold buttons. Between 1900 and 1901, $18 million in gold was produced in this area. Production in 1901 amounted to $17 million. Pride filled the district in a rainbow of glowing colors and world-famous dignitaries. Some of these figures were born and raised in the area. Others had been attracted to the region by its adventuresome reputation.

The towns of Victor and Cripple Creek had to be prepared for dignitaries of this sort. New buildings in Victor, built after the fire of 1899, had surpassed the elegance of those built after the 1896 fire in Cripple Creek. The Woods' Victor Hotel — still standing on Victor Avenue today — had become one of the most luxurious establishments in the whole region. The hotel's interior was decorated in the highest of Victorian fashion, and the exterior was constructed from the finest of "fire-proof" brick. All had been set on the solid foundation of red granite common to the land around Victor. Inside of the hotel, strips of yellow and green bunting lined the hallways. Semi-tropical plants filled every corner. Roulette, faro, and poker tables were from the grandest of European casinoes. Slot machines became attractive items in most of the gambling establishments. To this day, the hotel's excellent construction is still evident — a beautiful site waiting to be restored to its original historical elegance and grand position.

The Victor Hotel was not without its competitive counterpart in Cripple Creek. The National Hotel was a five-story structure of steel and brick. Built in 1900, some of its most attractive features were an electric elevator, private baths, and steam rooms. There were 150 guest rooms — more than enough space to handle and house elegantly dignitaries from all over the world.

Theodore Roosevelt, while campaigning for McKinley against William Jennings Bryan, came to the Cripple Creek-Victor Mining District in 1900. As candidate for vice-president, Roosevelt was after the votes of the Republicans for the silver standard and the anti-gold Democrats. Danny Sullivan, an associate of A.E. Carlton, had arranged Roosevelt's tour through the area. Coming up to the district on the Midland Terminal

The Victor Opera House staged a production of Lew Wallace's *Ben Hur,* and promoted the fights of Jack Dempsey.

Railroad, Roosevelt's first stop was the town of Victor. At the Victor Armory pro-Bryanites began to incite the people against Roosevelt. They started by shouting Roosevelt down as he spoke, then yelled obscenities and angry threats. They were not at all pleased with McKinley's plan to make gold a standard $20.67 an ounce. Soon the crowd transformed itself into a mob and attacked the podium of the former Rough Rider. Roosevelt's glasses were knocked from his face. Roosevelt struck out in defense. Sullivan and others surrounded Roosevelt and fought off the mob until they could return to the Midland Terminal train.

As would be expected, the residents of Cripple Creek treated Theodore Roosevelt much differently. They treated him as royalty. In fact, Roosevelt's stay in Cripple Creek was so successful that he accepted Attorney J. Maruice Finn's invitation to return to the town the following spring as Vice-President of the United States.

Finn immediately borrowed $30,000 from Carlton's First National Bank of Cripple Creek as soon as Teddy Roosevelt was on the train back to Colorado Springs. With the money, Finn set to work building a residence on Place Street and Pikes Peak Avenue worthy of a vice-presidential visit. When the mansion was finished, the five-story building had as its central core a magnificent spiral stairway which rose out of a flowing pond stocked with rainbow trout. Oak paneling lined the walls. Among its many rooms was a library, its bay windows decorated with leather seats. The outside features of the house included an observatory and Gothic turrets. Upon Roosevelt's arrival in the spring of 1901 as Vice-President, the mansion was lit up with more than one hundred Edison light bulbs.

Months later, J. Maurice Finn, after finding the upkeep of the house preposterous, called it a "monument to a fool" and closed it down. Everyone else called it "Finn's Folly."

On this same trip in 1901, Roosevelt came up to the district on the newly built Short Line Railroad. He stopped in Victor again — this time, he had a far more civil reception. Roosevelt was given the royal tour of the Gold Coin Mine by the Woods Brothers. Part of this tour included an 800-foot trip down the bricked-in main shaft. In one drift, Roosevelt was shown a rich vein of gold ore. Someone in the party asked Roosevelt to help dig some of the rich ore out of the mountain. Without hesitation, Roosevelt took off his jacket, grasped hold of a pick, and struck into the vein.

With the rising gold production figures and the growing number of visitors, the entire Cripple Creek-Victor Mining District was gaining respectability. A small English invasion was initiated when Winfield Stratton sold his Independence mine to a British firm in 1899. Through the next few years, an English managerial staff and their families, along with other English financial interests and English tourists, moved into the territory.

In the year 1901, the district saw greyhound racing at Spencer Penrose's Gillett race track and at the small track in Pinnacle Park. Cricket matches, English football, and boxing were also being held under English game rules. Restaurants in the area served fish and chips and English tea as exuberantly as they served fresh lobster and oysters. All of this took place at more than 9,000 feet above sea level.

Little London, on the plains below Pikes Peak, had to give what was due the mining district that not only gave it wealth but made it famous as a health resort. Without the attraction of gold, Colorado Springs would have never risen to the dream that General William J. Palmer had for his "exclusive" town.

Lodges and secret societies grew to prominence. Women's organizations increased to the point that their demands were not only heard but acted upon. These women's organizations forced gambling in Cripple Creek to be moved onto the second floors of the saloons. Streets were paved so that the women's ornate white dresses would not be soiled by mud and dust. Stone slabs were placed over dirt and wood sidewalks. The women would not have felt so haughty had they not had relatives and acquaintances from the East and Europe to entertain as visitors to the fashionable Cripple Creek-Victor Mining District. The women's next pursuit was temperance and the abolition of prostitution. However, these pleasures of the men were not so easily purged from the walks of Myers Avenue.

About this time, the Kansas Tornado — Carry Nation — decided that she must purge the mining district of all liquor and lascivious activities. Her

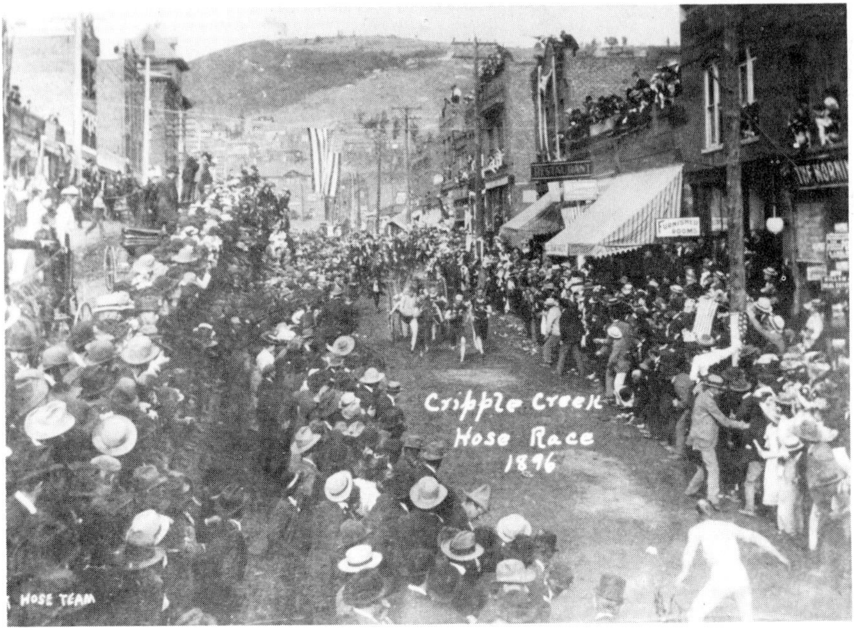

The people of Cripple Creek let down their aura of high society to frolic in some energetic competition.

approach had not been one of secrecy. Upon receiving word of her appearance, most of the saloons and parlour houses in the district were boarded up. Only Johnny Nolan and his Manitou Saloon were brave enough to fight the incoming storm. Carry Nation's six-foot figure entered the doorway of one of the plushest saloons in Cripple Creek. It did not take long for Nation to perceive that Nolan's Manitou was a den of vice. Her axe came from beneath her shawl and began sending to the devil all works of hell — including a painting above the bar of Venus rising from the sea (no doubt a copy of Botticelli's work, which even Savonarola could not destory).

The entire saloon turned into a classic barroom brawl, with Nation's army fighting Nolan's. Velvet curtains were pulled to the floor. Roulette wheels spun their last. Bottles of liquor were shattered on the beautifully lacquered floor. But the conflict was short. Ms. Nation was arrested and thrown in the county jail. At her court proceedings, she was fined $50 or thirty days in jail. Ms. Nation had no money. So, Mr. Nolan offered to pay Ms. Nation's fine if she were on the first train out of town.

Ms. Nation rode out immediately on the Midland Terminal.

The parlour houses had become some of the most aesthetic and elaborate buildings in the region. Such madames as Lola Livingston, Mattie Silks, Pearl De Vere, and Nell McCluskey nurtured parlour houses like the Old Homestead into pictures of luxury and elegance. Everything in these houses was dedicated to high-living. With teakwood tables, crystal lamps, Pier mirrors, Edison phonographs, excellent Victorian furniture, polished wood floors, and mahogany trimmings, European brothels were left far behind. Most of the utilities were electrical, and the indoor plumbing of the best parlours had hot and cold running water.

The zealous women's organizations had put a halt to street solicitation. So the girls of the parlour houses — in their fine Parisian gowns and straw-pleated hats — stood in bay windows and displayed their wares to interested customers much as the women in the bordellos in Germany.

May 19, 1901 saw the official opening of the district's only amusement park. Just west of Cameron, the Woods Brothers' Pinnacle Park was expected to be the entertainment spectacle of thousands. Mountain lions, bobcats, birds, and bears caught in the area were placed in the park's many cages. The park also had a large baseball field, a dancing pavilion, a race track, and a place for bands to play. Vaudeville acts were also staged. The park had been opened in 1900, but it proved so popular that it had to be enlarged for the 1901 season.

Pinnacle Park boasted that it had something for every member of the family. Electricity moved rides such as the merry-go-round. There was a shooting gallery, moving picture machines, and geological displays. Dining halls, saloons, and concession stands helped to make the days more en-

Spencer Penrose and Charles Tutt outside of their prospering real estate, surveying, and mine managing office. As members of the Socialites, their antics were far ahead of the gossip spread about them.

joyable. The entire complex took up a square mile of land in a wooded area which had mines no more than a stone's throw away.

Because of the increase in population and the miners' desire to live either in Victor or Cripple Creek, the interurban electric tram system was extended. This expansion included a loop around the business section of Cripple Creek — Bennett Avenue.

The miners could take an electric tram car to work at most of the major mines, whether they were in Poverty Gulch or atop Bull Hill. The interurban electric train was the highest tram in the world, rising to elevations of nearly 11,000 feet. Towns such as Midway grew from the trains that came up from Cripple Creek and Victor every half-hour. The Midway Saloon became a very popular stopover for miners going home from work. At the Midway Saloon, the miner could buy both liquor and women.

These increases in railroad lines — steam- or electric-powered — helped make the district even more productive. As the number of spurs from these railways multiplied to large as well as small individual claims, the rates of shipping ore decreased. This helped make small mines and the mining of low-grade ore more profitable.

In September 1901 the Short Line Railroad was operational but not quite finished. After Teddy Roosevelt rode the Short Line in 1901 and proclaimed that the scenery of the area "bankrupted the English language," the railway became an enormous tourist attraction. At its financial peak, the Short Line had 12 one-hundred-ton locomotives, 213 box cars, 69 coal cars, 12 club cars, and 3 cabooses. The railway was ten miles shorter than the Midland Terminal out of Colorado Springs. The year 1902 witnessed the passage of 36,000 tourists on the Short Line in a single month, and by the end of the summer more than 66,000 had ridden up to the Cripple Creek-Victor Mining District on this railway.

In essence, quality traveling accommodations and majestic scenery made the Cripple Creek-Victor Mining District a haven for tourists. The sights were innumerable — as they are today. Above Victor, and near Altman, a view of many of Colorado's Fourteeners in the Rocky Mountain Range could and still can be seen.

The Cripple Creek-Victor Mining District was an alluring vacation spot in the first decade of the Twentieth Century. It was wealthy in gold, built with the utmost of modern innovations. The people of the district were colorful, and life appeared to be constantly exciting.

But in 1902 the district's world reign as "King of Gold" began to crumble.

THE DEATH OF STRATTON MARKS THE DOWNHILL SLOPE

The Economic Mill in Requa Gulch was entirely run by mankind's newest energy source — electricity. The Cripple Creek District Railroad Company ran the highest interurban electric tram in the United States. On two 600-volt overhead lines, 40 passenger cars moved on six miles of sixty-pound steel rails using air brakes. But during this expansion, the district's oldest town site, Arequa, founded in 1889, was dying. The year was 1902.

Many of the district's other small communities were also fading. Altman, Mound City, Cameron, and Gillett saw marked declines in population, mainly because of the growth of the urban centers of Victor and Cripple Creek. But other factors worked against the small communities of the area.

Hard-rock mining was difficult and costly. When the underground water tables of the ancient volcanic area started draining into the mines and the cost of pumping the water out became excessive, smaller operations went out of business. Even electrical pumps became too expensive for the small mines to relieve their shafts of the great quantities of water flowing into them.

The milling of the gold ore was becoming over-priced and inefficient. The original chlorination mills had to be usurped by the more efficient cyanide and bromide mills. These mills used the McArthur-Forest Process which was modified in 1899 especially for the Cripple Creek-Victor area from a process patented in Scotland in 1867. The chlorination process had let too much of the gold be lost in slime, liquids, and tailings. Cyanide leaching was approaching 95 percent recovery.

Then, after dealing productively with the water and milling problems, many mining operations were confronted with a "gas" problem. This gas was mostly carbon dioxide and nitrogen. It would rise in the drifts and shafts and displace the already thin layer of air. This gas suffocated miners underground where the circulation of air was poor and not constantly pushed out by electrical pumps. Many miners, upon awakening in the morning, would look out into the distant Collegiate or Sangre de Cristos mountain ranges. If clouds clung to the high slopes of the mountains, they would not risk the lethal danger of gas in the mines. They had no firm conception of why the gas and low clouds were related, but their instincts had been correct. Low barometric pressure — the pressure that made conditions prime for low clouds — allowed this gas to emanate upwards from the remnants of the extinct volcano which had formed the district.

The Cornish miners had even come up with their own folklore concerning the gas of the mines. They believed in little creatures whom they called Tommyknockers. These leprechaun-like visions were friendly to men — if treated correctly — and only lived in working mines. Tommyknockers were supposed to warn miners of encroaching gas by blowing out their candle

One of the hoists in Stratton's "golden bank" — the Independence Mine.

lights. The Tommyknockers also warned the miners of possible cave-ins by dropping pebbles on the miners' helmets. Swearing at Tommyknockers could put a miner on their bad side, and the only way to get back into their graces was to walk through a particular drift apologizing aloud to them along the way.

Deep, hard rock mining acquired other exorbitant expenditures. Because of increased shipping, milling, and energy costs, the amount of mines in operation dropped from 475 to 83 by 1910. Between the years 1901 and 1908 the population of the entire district dwindled to 18,000. At $20.67 an ounce, it was no longer profitable for small mines to produce gold ore. Reluctantly, many mine owners closed their operations, leaving more gold in the ground than they believed they had ever taken out.

Gone were the days when the story of J.R. McKinnie, owner of the Mount Rosa Placer, could be believed. It was said that McKinnie merely kicked his boot into the loose soil below Victor and discovered a rich vein of "wire" gold. McKinnie supposedly took out enough gold from his placer claim in seventy-two hours to retire a wealthy man for the rest of his life. But by this time in the district, stories of this kind were considered little more than romantic dreams. Realism was now flowing into the subterranean passages of the mines. And everyone knew that J.R. McKinnie was rich, but he had not retired.

There were still many rich days to be had in the district. But the first phase of the Cripple Creek-Victor Gold Rush was over. Winfield Scott Stratton died on September 14, 1902. And, as if the whole district had etched a golden monument to this man, much of the district closed down.

With the rise of the power and energies of A.E. Carlton, a new golden era was on the way.

Outside of the Carltons' Colorado Trading and Transfer Company, the employees pose for a memory.

Winfield Scott Stratton had established the settlement of Stratton in 1900 for the sole purpose of uncovering the "Stem of the Goblet" of his "Wine Glass hypothesis." The core of this proposed wine-glass formation was believed to be buried under Iron Clad Hill. Stratton had been chiseling out shafts with various drifts all centered on this "great mother lode" beneath Iron Clad. These shafts were dug from the John A. Logan, the Orpha May, and the American Eagle mines. The Lady Stith mine served as a central point of crosscuts for the incredible system of drifts and shafts which were costing Stratton between $35,000 and $50,000 a month to work on.

The settlement of Stratton had been strictly a commercial enterprise. The steel-framed buildings housed the corporate offices and activities of his organization as well as the dormitories for the mine labor.

Nothing definite was ever proved by these "Stem of the Goblet" operations because all activity on the project ceased upon Stratton's death.

The work had been a great boon to the Cripple Creek-Victor Mining District because of Stratton's high expenditures on the project. After Stratton's death, his estate was devoted to the Myron Stratton Foundation in Colorado Springs, which had been set up to aid the elderly and orphans.

Stratton never forsook the district that had made him a king. While maintaining a small manor in Colorado Springs — which he had helped build when he was a carpenter — he also kept up a residence near his Independence Mine and Mill. In this building, it has been said that Stratton lived his life of real eccentricity. Yet, as with his Independence Mine, he maintained a certain degree of secrecy concerning the mistresses he arranged to visit him, about the profuse gambling and the raucous parties held in this house on Battle Mountain.

Eccentricity was the rule in Stratton's Battle Mountain residence. His Bohemian nature was as interesting and profuse as the gold he mined.

Still, the only outstanding luxury that Stratton continued for himself until his death in 1902 was his supply of custom-made leather boots. Stratton provided a fine income for life for his friend, Bob Schwarz, so that Bob could provide him with excellent leather-crafted boots whenever he wanted them.

With Stratton's death, the settlement of Stratton became little more than a storage area for his estate's records. Many of Stratton's mines stopped production at the time of his death. The "Midas of the Rockies" had owned almost one fifth of all the rich properties in the district. The area was dealt a severe economic blow upon his passing away. Besides all of his other mines, when Stratton's $50,000 per month operations stopped on Iron Clad Hill, there was a definite decline in the production of the district.

Winfield Scott Stratton had helped build Victor and the entire district. In his Independence Mine, an ore vein 18½-inch-thick had once been struck. The assay had a reading of $250 to several thousand dollars per ton. A section in the mine called the Bull Pen had produced ore which assayed at $450 per ton. The mine also contained another area which had petrified tree stumps 500 feet below the surface. They possessed an estimated four to five ounces of gold to the ton. Winfield Scott Stratton, who could be seen daily in his gray Stetson hat, creaseless pants, and high-heeled spotlessly polished boots, had been one of the outstanding figures of the district's first era.

GOLDEN GLIMMERS OF HOPE

Stratton's pursuit of the Stem of the Goblet had engaged nearly one-fifth of the mines in the district. The project included hundreds of rich ore veins and a multitude of large ore chutes. The Chicago-Cripple Creek Tunnel ran underground for more than 3,000 feet, connecting Gold and Globe hills with Iron Clad as its center.

Operations in the area did not totally cease despite the mounting deficits of hard-rock mining. The reign of A.E. Carlton was just beginning.

Anaconda, located in Squaw Gulch to the east and above the Economic Mill, proclaimed itself a healthy and growing town in 1901. Once named Barry, Anaconda was noted as a residential town. The gulch had acquired its name from the fact that it had served as superb protection for the squaws left behind by Utes who had gone to war against the Arapahoes. Its next historic claim was Texas Guinan.

The town of Anaconda had many large mines near its city limits and boasted that the Arno Mine, in its core, would soon be a highly productive venture. It was extremely proud of its secret organizations and lodges, such as the Red Men and Knights of Maccabees. But a destructive fire permanently maimed the town in 1904. Like Cripple Creek and Victor before it, Anaconda's water supply was low at the time the fire swept through the tinder homes. As its five saloons and club houses burned down and dynamite served only to demolish what was left of the town, much of its population decided to move permanently to Victor and Cripple Creek.

Although the population in Anaconda after 1904 dropped from 1,000 to 250, other towns in the district were more fortunate.

Dwellings of miners and their families in Anaconda.

A rich strike was made in 1904 on the W.P.H. claim on Iron Clad Hill. Sylvanite in rust and gold cubes was found. The cubes possessed a very high gold content. This discovery lead the leasees of the claim, Harrison and Seiver, to a vein of ore twenty inches wide. The property was leased by the Woods Investment Company, which received 30 percent of the take. The combined W.P.H. and Jerry Johnson claims were quickly promoted as being richer than the Isabella Group. When the figures were finally summed up, the Isabella Group had produced $15,700,000 worth of ore. The Jerry Johnson had produced only $5,000,000.

While the W.P.H. was being guarded on Iron Clad Hill by men hired by the Woods Brothers, the El Paso was preparing the richest shipment of gold ever to be recorded — one carload with a value of $600,000. Solid calaverite worth $58,000 to $60,000 a ton was brought out of the El Paso Mine near Beacon Hill in the same year as the discovery in the W.P.H. Mine.

In 1904, gold production was $14,500,000, up nearly $2,500,000 from the previous year. The year 1905 saw gold shipments exceed $15,600,000.

But the feeling of decline was spreading despite these few rich discoveries. Civic improvements were still initiated, especially in Cripple Creek. Streets were paved or regraded. The water system was repaired. And the highway system was expanded.

This Victor building was a lodge for various secret organizations, a Christian Science Hall, and union hall. The building is across the street from the *Victor Daily Record* where Lowell Thomas began his reporting career.

Paradise Alley, the "tenderloin" district of Victor, became even prouder of its red brick building, the Victor Opera House, when it served as the stage for many famous performances by the Barrymores, Lillian Russell, and Lew Wallace's *Ben Hur*.

Cripple Creek kept up its prosperous appearance, although its wealth was not what it had been. Existing houses were improved with porticoes and porches, but relatively few new buildings were constructed. The business district was still healthy in 1904, but the "tenderloin" district, Myers Avenue, was in obvious decline. Many of the women of the parlour houses were showing their age. Now that their beauty and other attractions had been lost, the women's organizations' battles against the parlour houses had begun to overtake them.

LABOR WAR — AGAIN

The Cripple Creek-Victor Mining District since the Battle of Bull Hill in 1893-4 had become an overwhelmingly union camp. Eight unions governed most of the labor market. By 1903, another strike was brewing. This one began in the Colorado City ore mills. The first phase of the strike was leveled mainly against the Standard and Telluride mills and did not directly concern the Cripple Creek-Victor District.

This was soon to change. Early in March, the labor force of the district was asked to walk out of the mines in sympathy with the union members of the mill. The walkout was rejected by a 70 percent negative vote.

Bill Haywood, Executive Secretary of the Western Federation of Miners, was not satisfied with the reaction of the Cripple Creek-Victor miners. The miners did not care to upset their $3.40 per eight-hour day pay schedule since it was the highest in the Rocky Mountain region. Haywood had to find another way to incite the miners to strike.

A campaign of terror was unleashed on the mines and nonunion men. The damage from these activities was relatively small compared with what was to come, but it was enough to get the secretary of the Mine Owners Association irritated. Secretary Clarence Hamlin was not fond of the union in the district, nor was he persuaded of their firm power over the miners by their acts of terrorism. Hamlin decided to start issuing permits to work in the mines to initiate his plans aimed at clearing the union out of the district. Permits were issued depending on whether the miner belonged to the WFM. If he did, he was denied a permit.

This action by the MOA (Mine Owners Association) pleased Haywood because it irritated the union miners. He knew that his next request for a strike would be met with enthusiasm. On March 22, 1903, Haywood succeeded in getting 1,100 men out of the mines in support of the mill workers' walkout.

A dormant mine within the city limits of Altman. The backside of Pikes Peak is in the distance.

Governor Peabody of Colorado had been elected to office owing much to the generous assistance of the mine owners of the Cripple Creek-Victor Mining District. When trouble began in the district, reminiscent of 1893, Peabody reacted. Adjutant General Sherman Bell, one of Vice-President Theodore Roosevelt's select Rough Riders, was sent with the Colorado State militia to mediate the matter.

Scabs were already working in the Golden Cycle, Vindicator, Ajax, Gold Coin, El Paso, Tornado, Elkton, Strong, and Thompson mines. The Mine Owners Association, which had originally been formed to combat high-grading, was not going to let a strike similar to 1893's cripple their production again.

Tension and terrorism were growing. When General Bell came into the situation, he thought he was in the middle of war. He placed troops upon advantageous hills, equipping them with artillery and searchlights. Then a bull pen was constructed in Goldfield. The town of Goldfield became the militia's headquarters.

Bell believed the *Victor Record* was the main vocal piece for the strikers and had to be silenced. He decided to stage a raid on the newspaper. Forty-five armed men rushed into the *Record's* office, wrecking whatever got in their way. The editor, George Kyner, and four other men were arrested and thrown into the bull pen in Goldfield. Four WFM officers whom Bell had imprisoned earlier were already in the pen.

Shortly thereafter, Emma Langdon and Barbara Frietchie barricaded themselves in the *Record's* damaged office. The two women set to work as best they could in printing the next edition of the newspaper, then distributed it about the district themselves.

Meanwhile, the union had put blacklists into action. The strike was beginning to resemble the one of 1893.

Judge William Seeds in Cripple Creek ordered the release of the prisoners in Bell's Goldfield bull pen. He emphatically reminded Bell that no official martial law had been called into effect. Bell refused the order. Judge Seeds wrote out writs of habeas corpus. Bell answered the writs by setting up Gatling guns on Bennett Avenue and placing sharpshooters on the National Hotel. Then, with troops surrounding his prisoners, he produced them for inspection in the town of Cripple Creek.

Seeds held firm on his judgment to release the prisoners. After Bell's show of force, he returned to Goldfield with his captives.

A few days later, General Bell released all of his prisoners. The strike had begun to calm down. Hamlin's issuing of permits had made the ranks of the WFM quite thin. Bell's troops left the district. But Haywood had not conceded to the Mine Owners Association.

The acts of violence were bolstered. Trains were wrecked. Scabs were beaten. Dynamite was loose amongst the union miners once more.

August 10 came around and 3,500 miners walked off their jobs. The strike's tension had increased again to the point that both mine owners and miners were afraid of renewed destruction.

At about this time, Harry Orchard entered the scene. Originally a cheesemaker, Harry had become very dynamite-happy during his days of high-grading in the Vindicator Mine. Harry admired Bill Haywood and wanted to help him out as much as he could. Harry figured if he blew up a few scabs it would help the strike along. Orchard came up with a plan. Moving a carload of dynamite onto a level of the Vindicator Mine on which he believed the scabs to be working, he set up an ingenious contraption. Attaching a pistol to the cage of the hoist on the seventh floor, the hammer was cocked. When the gate was opened the pistol would fire into the dynamite caps and explode the cache.

The rig was set. But Harry became very disappointed when nothing explosive happened within the Vindicator for days. Then word came that Superintendent Charles McCormick and Melvin Beck had been blown apart on the Vindicator's seventh level.

Harry had merely miscalculated the levels. The explosion of November 21 had missed the fifteen or so scabs working in the mine because Harry's rig was on the seventh level while the miners were on the sixth. Superintendent McCormick and Assistant Beck were slaughtered while on an inspection tour of the mine's seventh level.

The State Militia came riding back, and martial law was proclaimed throughout the district. The mine owners were not taking chances over this renewed alarm. They hired professional gunmen by the hordes to search out WFM men. Gangs started ruling the hillsides. Everyone seemed to be fighting for a different cause.

Federal troops were urgently requested by the Colorado governor. Soon there were so many bands of men roaming the hillsides that no one was able to calculate how many factions there were in the strike. Finally, General Bell, after building another bull pen in Goldfield and filling it with hundreds of union men, made the miners surrender. They had to forfeit their arms and obey Bell's law.

Things were obviously not going very well for Haywood. He had spent more than $100,000 in a propaganda campaign against the Mine Owners Association, Clarence Hamlin, and General Bell. Yet the tide would not turn his way. In February 1904 the strike was officially called off.

Haywood did not fair too well at the WFM conference in Denver that year, and Harry Orchard was there to witness his humiliation. Orchard was positive he could help Haywood out.

In June 1904, Orchard's dynamite fever had infected him again. He had set up a rig on June 5 under the depot of the Florence and Cripple Creek Railroad near the town of Independence. By pulling on a cord he had secured in the Delmonico ore-house — out of sight of curious eyes — he would topple vials of acid which he had hanging over the blasting caps he had set over two mighty charges of dynamite. At 2 a.m. on June 6, the night shift was getting off from A.E. Carlton's Findley Mine. As the miners gathered on the platform and the train pulled up to the station, Orchard set off his rig.

Twenty-seven men had gathered on the platform when the two explosions shattered the depot into wooden toothpicks. The concussion of the two cases of dynamite exploding had nearly blown the train off the tracks.

The emergency call was relayed. Carlton and others had gotten up to the area before 4 a.m. Already the searchlights were sweeping the area for bodies. In his search, Bert Carlton located a man's leg with the boot still on it. Thirteen of the twenty-seven men were found in bits and pieces. The others were seriously wounded and only saved by the amputation of limbs.

Harry Orchard escaped to Wyoming. He disappeared for a few years but was later arrested for blowing up ex-Governor Steunenberg of Idaho. Orchard was sentenced to life imprisonment and caught a fervor of "religion" just before he died on April 13, 1954.

When the sun rose June 6, Hamlin had stood himself up in a wagon on a lot below the Gold Coin Mine to speak to a gathering crowd. Almost immediately Hamlin was heckled. Then a shot was fired. The horses tied to the wagon in which he was standing bolted. Hamlin was thrown to the ground. Everyone thought Hamlin had been shot. A massive brawl started as men tried to disperse in different directions. Three men were injured. Two died.

The anti-union people were seething. The State Militia had surrounded the Union Hall in Victor where a group of union men had barricaded themselves. General Bell hoped to bargain with the WFM members inside the hall. They would not listen to him. Bell ordered the militia to fire a fusillade of bullets into the hall. After many of the WFM members had been wounded, they surrendered. No sooner had the union men gotten out of the hall than the anti-union men stormed in and tore up everything they could find.

The contagion of the mob spread. Throughout the day, anti-union gangs were destroying every union hall and store they could get to in the district.

On June 7, deportations of union men began. 25 WFM members were shipped to the Kansas border. Seventy-five more were taken down to the deserts of New Mexico. When all of the deportations had ceased, 225 union men had been shipped out. Soon their families would follow.

A section of modern-day Cripple Creek.

After the union had been toppled from power, changes in the Cripple Creek-Victor Mining District came rapidly. The whole camp became a "company town." Many of the miners who had remained loyal to the WFM but had not vocalized their preferences left the district and moved on to the new gold discoveries in Tonapah and Goldfield, Nevada.

Bill Haywood, after spending more than $100,000 in support of the strike and on propaganda against Bell and the mine owners, left the district. He founded the IWW and then moved to Russia. In Russia, he worked for the Communist Party and died a hero.

The district was free of unionized labor, and it would never be the same. The violence of the strike had caused great fear. Investors added this labor trouble to the reports of dwindling gold veins and increasing hard-rock mining costs and decided to find other places for their money. The Cripple Creek-Victor Mining District was entering a lean period. Another strong leader such as Winfield Scott Stratton was desperately needed.

Some people had hoped that Jimmie Burns would take over Stratton's position in the region. But Burns had been a friend of the WFM. During the strike, Bell had to close the Portland Mine. Burns retaliated with a suit against Peabody, Bell, and the entire Mine Owners Association.

The directors of the Portland voted Burns out of the office of president. At first, Burns was despondent and lost. Jimmie Burns could see nothing for himself but to open up his residence on "Millionaire's Row" — Wood Avenue in Colorado Springs. In no time, the life of Little London began to suit his taste and his wealth.

He started to compete in fashion with Spec Penrose. Whiskey became one of his favorite pastimes. He acquired all the best that Colorado Springs had to offer and more. Burns married Olivia Belle Parker and became a respected citizen of Little London. His wealth continued to flow in from the Portland and his other gold interests, but he would never be the eccentric and outstanding figure of the Cripple Creek-Victor Mining District as Winfield Scott Stratton had been.

A minor part of the Free Coinage of Silver movement of the WFM — though still invaluably intriguing — was a man called Joseph Lesher. Lesher wanted to set an example for the pro-silver people. So he designed a coin with eight sides to be minted out of silver. The people of Victor accepted Lesher's design, then put the coin into print in 1900. The coin was minted in Denver. This silver coin was seen by Teddy Roosevelt and William Jennings Bryan on their tours through the district for the presidential campaign. The coin, because it did not have federal backing, was used much like the scrip of the world wars. Agents of the Federal Treasury Department confiscated Lesher's minting plates after only a few of the coins were placed in circulation. After many court battles, the plates were returned to Lesher. He began issuing his coins for use in stores as credit. Only a few years and variations of the coin were minted before the Free Silver Coinage movement faded. Today, the Lesher dollar is worth more to some collectors than a $20 gold piece minted by the United States government.

The remnants of the Midway Saloon where A.E. Carlton almost met an early demise at the hands of a mule skinner.

THE HEARTBEAT OF CRIPPLE CREEK

Albert Eugene Carlton was in Colorado Springs in 1891 when Winfield Scott Stratton was trying to unload one-half interest in his Independence Mine so that he could obtain some working capital. The Carltons had moved to Colorado Springs so that Albert might have a chance of recovery from tuberculosis. Stratton had offered Carlton half interest in the Independence for $500. Carlton turned down Stratton's offer because he did not think he had much time to live.

Bert Carlton thought that he would be dead within six months, thus he had no use for an unproven gold mine. Bert was born in Warren, Illinois, in 1866. In 1889, the entire Carlton family moved to Colorado Springs. By the fall of 1893, Bert had recovered from his struggle with tuberculosis. He had lost the use of one of his lungs but was strong enough in 1893 to travel up Ute Pass with his brother to see what all the excitement was in Cripple Creek.

The Carlton Brothers — Albert and Leslie — had been enthralled with what they had seen. Returning to Colorado Springs, they encouraged their father enough in their ideas for the area on Pikes Peak's southwestern slope that they both received a $10,000 stake. Bert had come to realize that the key to the crown of power in the district was transportation. The Carltons started almost immediately on their plan to conquer the area.

They began with a small business which supplied firewood and coal to many of the establishments and mines in the district. Bert was the promoter and the salesman. Leslie was the workhorse. Bert was able to talk anyone into almost anything. Leslie easily made Bert's word reality.

One of the Carltons' first accomplishments was the formation of a monopoly of coal delivery with the Midland Terminal. Their next accomplishment was an exclusive contract for the hauling of gold ore from the mines to the Midland Terminal. Prosperity of a moderate sort let them build an office at Fifth and Bennett avenues. They called the establishment the Colorado Trading and Transfer Company.

Bert had another plan to secure their business prospects. He traveled to Pueblo to talk with John C. Osgood. Osgood was the head of the Colorado Fuel and Iron Company. Bert figured that if he could get Osgood to agree on a deal wherein C.F. and I. coal was the only shipment delivered to the Cripple Creek area, then a lucrative monopoly could be formed between the Midland Terminal and the Colorado Fuel and Iron Company with the Carltons as the exclusive mediators.

The monopoly worked as Bert had planned. In fact, the business deal worked so well that Bert was able to buy the First National Bank of Cripple Creek by the year 1898. At the age of 33, Bert was worth more than $500,000, and his prospects were looking better than ever.

A.E. Carlton was an interesting man. His six-foot frame was thin yet strong. He was extremely appealing to women — so much so that he had two fiancees and one wife in 1899. Although he was a member of the Socialites, he drank very little, and his only gambling was in business dealings. Women, since he idealized them so much, were his major weakness. But they never got in the way of business. His major hindrances were his one useless lung and chronic ulcers.

Carlton had women problems, but he solved all of them before he was married to Ethel Frizzell in December 1901. Bert and Ethel had gone to New York City to be married. Upon their return to Cripple Creek, Lew Dockstader's minstrel band signaled the beginning of a grand celebration in the town. After their initial greeting, the Carltons were carted off to the National Hotel where a party ensued until the next morning. Thirty-seven cases of champagne were consumed, and it was said that for every glass filled another rose was thrown into the middle of the ballroom. The clean-up crew of the hotel found scores of empty champagne bottles and a ballroom full of roses.

As soon as the celebration had ended Bert was back to work. A large

From Stratton's John A. Logan Mine, a distant view of the Cresson Mine as it is today.

apartment above his bank served as the Carltons' home and business center. The next fourteen years were extremely robust and profitable for the Carltons. Their entertainments were spectacular. They brought the elegant crowd — the Upper Tens of Cripple Creek — to the area. Tuxedos and long gowns were seen nightly on the streets.

Bert increased his property holdings throughout the district. Mines were bought and sold as playthings on a Monopoly board. Ethel was gracious and exuberant about entertaining guests. Twenty-person lunches were not rare. Her greatest problem was keeping an excellent cook in the apartment. At the altitude of 9,500 feet, few culinary dishes turned out as they would have on the continent of Europe. Still, fresh lobster dinners were as prominent as stock battles at the Carltons' dinner table.

Early in the first decade of the new century, Bert Carlton was growing to a power which equaled the status of Winfield Stratton, the Woods Brothers, and Jimmie Burns. His properties had increased radically. His bank's loan policy was one of generosity, and his credit terms were fair. But the best advantage he possessed in giving out such credit was the inside knowledge he received on all the mines. He knew what mine was producing, what mine was floundering, and what mine was being mismanaged. Bert Carlton could not have been in a better position to start an empire.

From the inception of his businesses, Carlton had many friends. Being one of the Socialites kept him in contact with most of the millionaires of the district. Carlton was a charter member of Lodge No. 316 of the Elks in Cripple Creek. Members included such notables as Jimmie Doyle (part owner of the Portland Mine), J. Maurice Finn, Danny Sullivan, and Spec Penrose.

The Victor Mine southwest of Altman.

Bert Carlton, however, had a powerful foe. This was a man named John T. Milliken. Milliken had bought such rich properties as the Anaconda and Golden Cycle from Dave Moffat when he sold out of the Cripple Creek-Victor Mining District. At the time Milliken was buying into the area, Carlton was enlarging his holdings with the prosperous Doctor Jack Pot and Findley mines.

Milliken would soon become Carlton's only obstacle to forming one huge consolidation of mines, mills, and related businesses throughout the entire district.

Bert and Leslie believed that an integration of services and mines would lead to greater profitable production. They began on their plan to unify the district. While trying to accomplish this, Bert was not above a touch of business subversion to get what he wanted.

During one of the transportation battles in the area, Bert had a few carloads of coal dumped in the way of the workers on the High Line interurban. The owner of the interurban, L.D. Ross, was hoping to construct a loop through the business district of Cripple Creek. The idea had also occurred to Carlton. The battle was on. In order to delay Ross' work, Carlton had coal dumped on Bennett Avenue directly on the construction route. This stopped the workers on the High Line long enough for Carlton's men to get the workers drunk and off of the job. His next move was to jump ahead with his project devoted to the business loop.

The war between the Midland Terminal and the Short Line also involved Bert Carlton. This time, Bert's foe was Jimmie Burns. But Bert won that battle, too. Burns had to sell the Short Line to the Colorado and Southern, which in turn leased it to the Midland Terminal and Bert Carlton.

All the way around, A.E. Carlton was an industrious and amazing figure. When the mines began flooding at the new lower levels, Carlton saved a tunnel project in Gatch Park from absolute disaster. From 1907 to 1910, Carlton supervised the drilling of the Roosevelt Tunnel. When the tunnel was completed, many of the mines began producing profitably once more.

But there was always a battle for Bert Carlton to fight. The fight was what kept him alive. When the Woods Brothers overextended their operations and their $45 million empire collapsed, John T. Milliken bought up their huge consolidation. United Gold Mines Company had more than a hundred mines under its power, but the Woods Brothers had not foreseen the energy of their competition. Milliken was quick to snap up what was left of the Woods empire and rebuild it into his. Milliken formed the Golden Cycle Mill Company. This enormous company had all the properties of the United Gold Mines Company, an ore-reduction mill in Colorado City, and a refining mill in Colorado Springs. Milliken was well on his way to owning the Cripple Creek-Victor Mining District.

The war between the corporate minds was again in full gear. Carlton tried blocking all of Milliken's moves, and Milliken did the same for Carlton. When Bert tried to raise the prices for the shipment of coal on the Midland Terminal, Milliken announced plans for an aerial tram that would provide the most direct and cheapest of all transportation routes. This aerial tram had been proposed to go straight down the mountains to Colorado Springs and Canon City, instead of curving about in gullies and canyons down the steep grades of the district. The higher prices were not posted for the Midland Terminal. Thus Milliken's aerial tram never materialized.

Finally the war became a single battle. Bert Carlton had rounded up some of his millionaire buddies and formed a syndicate to buy out Milliken. The $200,000-a-month copper wealth of Spec Penrose came into play on behalf of Carlton. Proxy battles were fierce. Directors and stockholders were at each other's throats. Carlton owned large amounts of stock in Milliken's company, and Milliken owned great chunks in Carlton's. Lawyers, contracts, meetings, and offers pierced the air like bullets.

The gallows frame of the Solomon Mine on the edge of Midway.

In February 1915 came the announcement of a deal between Carlton and Milliken. The Golden Cycle Mill in Colorado City was sold to A.J. Zang (owner of the Vindicator Mine and Zang's Brewery in Denver) for $1 million. Carlton, Penrose, and E.P. Shove bought out the Golden Cycle Company and all related properties for $6 million.

Carlton issued stock immediately. The initial $1 per share shot up, and in two months a dividend for the stockholders was declared.

Through the same syndicate of Carlton, Penrose, Shove, Zang, MacNeill, Roelefs, Burbridge, and Boettcher, the Cresson Mine was bought for $3 million. Stock was issued in 1915 for $3 a share. By 1919, Cresson stock was selling for $13.80 a share.

Bert and Ethel Carlton moved to Colorado Springs. Naturally, they moved onto Wood Avenue — across the street from Jimmie Burns. But the business dealings in Cripple Creek were still going on.

Bert had sold the Colorado Trading and Transfer Company in 1898 to buy the First National Bank of Cripple Creek. Now Bert bought back his old company. He also bought the controlling shares of the Midland Terminal from H. Blackmer in Paris. In 1922, Bert bought the Vindicator from Zang. By the mid-1920s, practically everything in the Cripple Creek-Victor Mining District was Bert, Ethel, and Leslie Carlton's.

On July 1, 1915, the Independence Mine was bought from the Stratton Independence, Limited of London by the Portland estate. In the old English custom, a section of soil, a key, and a piece of ore sealed the bargain. So the old Independence was American once again.

Carlton helped form the Holly Sugar Corporation, then moved into Spec Penrose's Broadmoor Hotel. The Portland estate was bought by Carlton's Golden Cycle Company in 1933. However, Bert Carlton was not around to see this purchase. September 7, 1931, was the day of his death — he was sixty-five.

Leslie Carlton took over as head of the corporation and ruled in a similar manner as his brother until his death in 1938. Then the latent powerhouse, Ethel Carlton, resumed control of the empire. She had the Carlton Tunnel bored through six and one-quarter miles of granite to drain the district down to the 6,900-foot level.

Ethel was extremely industrious. In 1951, she had the Carlton Mill built. This mill still stands on Highway 67, just below the Elkton and Cresson mines. This cyanide mill cost $2,000,000. It could profitably process ore in 1951 which assayed as low as $3 a ton.

But fortune was not on Ethel Carlton's side as it had been for Bert. Wars kept depleting the labor force and raising the cost of gold mining. And if these factors were not discouragement enough, the War Production Boarding Ruling L-208 most assuredly was. This ruling halted most of the mining of gold in the area between the years 1941 and 1945.

It is ironic to note that many of the miners who survived the battles on Bull Hill, two of Altman's secessions from the country, the deportations of

1904, and the gases and floodings of the mines were killed in the armed services during World Wars I and II.

Yet, the Carlton corporation generally prospered because of its varied business interests. The Carlton operations, at the time of Bert's death, had a combined production account of $227 million.

HEAVEN IN THE CRESSON

The Cresson Mine had originally been located around the year 1895. Little had been done with it since the claim was staked. All the way around, the Cresson had been unfortunate. Jesse Waters of the Midland Terminal had refused to run a spur to the mine to aid its production because of the rough terrain the claim had been located in. And, as far as anyone knew, the Cresson had not produced any ore worth speaking of. So for nearly twenty years the Cresson was considered a low-paying mine.

J.R. and Eugene Harbeck had come by the mine in 1910. After buying the property, their first manager came up with nothing but debts for them. The Harbecks went to Bert Carlton for a recommendation of a new manager. Bert supplied them with Dick Roelefs.

Roelefs was an innovator in hard-rock mining. Many were afraid of his odd techniques. But Bert Carlton had faith in him, so the Harbecks accepted Roelefs.

Using the pillar method, which dug laterals away from the main drift, Roelefs struck acres and acres of low-grade ore in the Cresson. The Harbecks, along with most everyone else, were amazed. The ore was valuing out at $15 per ton.

Deep inside of the mine, Roelefs' stopes were huge. Entire houses could be placed in these stopes. But there was still the problem of transporting the ore down the mountain. The Midland was not available, so Roelefs engineered another of his eccentric tricks. He devised an aerial form of transportation which ran a mile down the mountain to the town of Elkton. Under Roelef's management, the Cresson became a healthy producer.

Dick Roelefs soon acquired the title of "Miracle Miner." But his $15 per ton discoveries were meager compared with what he was to find.

Late one cold night in 1914, Roelefs awakened Hildreth Frost and asked him for his assistance. He needed a respectable attorney as a witness. Roelefs knew Frost's word as a lawyer would be accepted by the rest of the community. Frost asked Roelefs what all the commotion was about. Roelefs would not tell him. The mystery excited Frost, and he agreed to meet Roelefs the same night. Roelefs had also persuaded Ed De LaVergne to meet him. When the three men got together, they took the Low Line to Elkton. From there, they hiked up the mountain to the Cresson Mine.

Roelefs took Frost and De LaVergne down into the mine. On the twelfth floor, he walked the two men around until they were thoroughly confused about where they were. Then, Roelefs led them to a steel-barricaded stope. He rapped a special signal on the steel door. There was a pause.

The door opened. Six guns pointed out at Roelefs, Frost, and De LaVergne. One of the surprised men was about to protest when he saw the reason for the guards. Magnesium flares lit up the huge cavern, called a "vug," and the stars seemed to be sparkling at close range inside of the mine. The great geode formation was a devastating sight. Millions upon millions of quartz and gold crystals lined the walls of the vug. Sylvanite and calaverite flakes clung to the quartz-bearing rock. The whole scene appeared to be the work of an artist. The rounded-out section of rock was forty feet high, twenty feet long, and fifteen feet wide. Neither Roelefs, Frost, De LaVergne, nor the six guards had ever seen anything like it.

The men stared and stared and knew that the Elysium Fields could be no more heavenly.

But the beautiful gold-and-crystal formations had to be destroyed. When the mining was all over, 1,400 sacks of crystals had been gathered up. Within four weeks of work, the gold from the vug was totaled at $1,200,000. One shipment weighed some 150 tons and brought in $468,637.29.

The Harbecks were elated. The stockholders were even more titillated. The Harbecks and the stockholders convened and agreed to buy the Miracle Miner a new automobile — a $4,500 White. Roelefs was a mining engineer wizard. He could do anything he wanted with the enormous hard-rock machinery for mining. But Roelefs was never able to learn how to drive his new car.

Roelefs soon became a major stockholder in the Cresson. In a few short years, the Cresson produced $45,000,000 in gold. When Roelefs retired, he moved to New York and became one of the most fanciful men in the city.

There had been no discovery like it throughout the whole district. The only spectacle that may have come close was the "Christmas Tree" of the Jo Dandy Mine. The miners in the Jo Dandy had uncovered a petrified pine tree some 600 feet below the surface. Petrified and mineralized, the hardened formation sparkled with quartz and gold crystals. The pine needles were said to be brilliant quartz crystals with an inner lining of gold. Of all the photographs of the district, no known picture was taken of this sparkling Christmas Tree or the Cresson vug.

DECLINE

Not all was fortunate during the second decade of the new century. An evident decline was creeping through the whole Cripple Creek-Victor Mining District. The competition between the towns of Victor and Cripple Creek subsided. The battle for supremacy became a battle for survival.

The years 1908-1919 saw most of the lesser communities disappear. Automobiles entered the picture, but they saw very little use other than in moonshining. World War I was taking most of the mining labor, and hard-rock mining costs were rising rapidly. The Golden Cycle Mill was taking on most of the ore that trickled out of the district. But the golden era seemed history.

The Cameron Mine north of Pinnacle Park and west of the town of Cameron.

The year 1933 saw a mild gold rush. The price of gold reached $35 an ounce. The district suddenly had people again. The old railroad beds were ripped out of the ground and replaced with roads for automobiles and trucks. All of the houses and stores were full once more. Repair and renovation was occurring all over. But there was no evident expansion.

The gold rush dwindled with the onslaught of World War II when mine labor was once again taken for the European fronts. During this barren period, many structures were torn down because of the high tax base on mining property in Colorado. The parlour houses of Myers Avenue had disappeared, with the exception of Hazel Vernon's Old Homestead. The Baltimore Hotel in Victor, where Groucho Marx and his brothers had once stayed, was knocked down. The National Hotel in Cripple Creek was no longer in use and was destined to be leveled.

The Carlton Tunnel had been completed in 1941, but there were few miners left to take advantage of Ethel Carlton's labor. However, in 1946, mining began to grow again. New strikes were even made.

The Golden Cycle Mill in Colorado Springs closed in 1949. And on February 6, 1949, the Midland Terminal made its last run.

There was still some hope for the area. The Carlton Mill, between Cripple Creek and Victor, was opened in May 1951. Structures were still being torn down in Cripple Creek and Victor, but the prospect of a new mill within the boundaries of the district paved the way for new mining.

Texasgulf recently signed a working contract with the Golden Cycle Corporation. The Carlton Mill will be revitalized, and its 1,000-ton-per-day ore-processing capacity restored by Stearns-Roger Engineering, Inc. Many of the old mines will be renovated. With the prospect of gold soaring from $400 to $1,000 an ounce, the value of the district will be realized again.

THE WINE-GLASS HYPOTHESIS

When Winfield Scott Stratton returned from his trip to England in 1899, he was no longer in possession of his great Independence Mine. But he had 666,666 shares of stock to sell in a newly formed English company which would eventually make him ten million dollars. On June 15, 1901, he had an announcement to make to the world.

The Stratton company was going after the "Stem of the Goblet" in the Cripple Creek-Victor Mining District.

The "Wine Glass," "Bowl of Gold," or "Stem of the Goblet" were all the same name for a hypothesis about the Cripple Creek-Victor Mining District. Through a thorough study of the area, Stratton had combined his geological knowledge with the age-old desire of man to discover the "great mother lode." Stratton believed that all of the major veins of the district flowed together in a sort of wine-glass structure. This caldera was a highly concentrated formation of tellurides stretching deep into the earth.

Stratton knew that the mining area sat upon the site of an ancient volcano, but its original shape had been eroded down into the form of a highly uneven plateau. Further geological study showed that the granite base of the area had been covered with gneisses, schists, and sandstones during the Paleozoic, Mesozoic, and Teritary eras. Then, volcanic eruptions violently cleaved the pink granite and pushed up the lava flow into a huge volcanic cone.

When the lava flow touched the cool weather and air, it exploded into a wide variety of hardened rock from crystalline conglomerates to fine sand. After the first set of vulcanisms had settled, a second wave of explosive activity burst through the hardened material. The breccia, a conglomerate of fine-grained sand and jagged fragments which had been pushed through the granite base, was then permeated by phonolite. Soon, a full maze of geological fissures filled the granite area.

Minor pockets of volcanic eruptions began bubbling up throughout the district. Stratton believed the major crater was just southwest of Bull Hill. Minor craters formed around this major one. Beacon Hill, Globe Hill, Mineral Hill, Copper Mountain, and Ryolite Mountain were all believed to be subroots of this major caldera.

Graves in the Victor Cemetery.

Gases and liquids flowing upward met with the cold air and rains of the area. The breccia or "porphyry" condensed or shattered under these conditions. Gold salts flowed up from the caldera in alkaline telluride solutions. Silver, quartz, copper, pyrite, and fluorspar also came to the surface during this second wave of eruptions. The craters were from 4,000 feet to incalculable depths. The "pipes" of molten rock formed veins in the cracks and crevices of the granite base. Here the gold salts precipitated into the porous breccia. Today, this breccia is generally found in red, purple, or gray colors. Exposed to the atmosphere, the breccia becomes a light-grayish yellow. The gold is most often found in this grayish-yellow rock.

The gold ores of the Cripple Creek-Victor Mining District are unique. They are generally found in the composites of calaverite and sylvanite tellurides. In their basic form, they do not appear as yellow gold. Their crystalline formation has more of a silvery than auric appearance. Oxidized gold dust near the surface is often brown in color.

Tellurides contain tellerium, gold, galena, and some silver — 3 to 13 percent. Some gold is found in the area in the form of sulphides. This formation is called a tetrahedrite and is often described as "gray-copper" in color. Gold was rarely ever found in the preconceived notion of a nugget. There was some flake, wire, and speck gold discovered, but placer finds were extremely rare. Altogether, only 380 ounces — out of the 22 million unearthed — were found in the free, nugget-like form.

This is why the call had gone out, in the very beginning of the first Gold Rush, that the tenderfeet were finding the gold while the prospectors were looking for it where it should have been. The silvery gold of the Cripple Creek-Victor Mining District was not following the same geological patterns that the old prospectors had been accustomed to searching for. The gold lodes were difficult to find, and they were dissimilar to the strikes in California, Alaska, and South Africa. Even the strikes of Central City, Colorado, were very much different from the gold ore of the tellurides of the Cripple Creek-Victor area. There were no major lodes exposed at the surface in the Cripple Creek-Victor area. The only way to discover a vein was to dig into the soil. Many of the most productive veins, such as that of the Cresson, were not found until much work had been performed on the mines.

Stratton had charted many of the fissures which the tellurides had precipitated and leached into. He grouped them into five regions. There was the Independence-Portland region where the veins had a northern flow. The Gold Coin — Dead Pine region had a northern flow. The Elkton-Raven region had a northeastern pattern while the Lillie-Vindicator had a northwestern. And the Mary Mckinney had a northeastern slope. Using these graphs and figures of the producing mines, Stratton used his modified wine-glass hypothesis to arrive at a plan for the investment of his millions of dollars.

Buying up 600 acres of mines and claims with more than $4 million, he set to work with another $3 million in proving his hypothesis correct. The American Eagles, Orpha May, and John A. Logan were at the center of his plan. All of these surrounded an Iron Clad Hill claim called the Lady Stith. This mine had a three-compartment shaft called the Plymouth Rock. The drifts of this mine were braced with eighteen-foot timbers, which would have allowed the mining to go below 5,000 feet. The array of drifts, crosscuts, and air tunnels grew quickly. The Chicago-Cripple Creek Tunnel was bored out for 3,000 feet, connecting underground Globe Hill with Gold

Hill for the express purposes of prospecting and allowing air circulation. The network of tunnels and drifts were spectacular. Stratton was spending more than one million a year to develop his hypothesis.

Two disasters struck the project before it could be completed. Water — the district's great recurrent problem — began flooding the lower levels. The base of the extinct volcano of the district is granite, so the region is generally impermeable for an outflow of water. But the phonolite, breccia, and fissures make the district a rocky sponge for rain and spring run-offs. However, the surface of the district is on the whole very dry. When water was hit in the lower levels of the project, the idea of another drainage tunnel came into play. But Stratton died in 1902, and the entire workings were abandoned because no one else had the money or the courage to pursue such a hypothesis.

Even John Hays Hammond of the Venture Company of England, who had warned investors that Stratton's Independence was petering out, thought that Stratton had bought some of the best land in the district for his project. When the rest of the world heard of Stratton's wine-glass plan, they compared him with Cecil Rhodes of Witwatersrand, South Africa. Some thought him crazy, pursuing an ancient myth of mankind that everyone knew was not true but had always been secretly wished for — like the sight of a unicorn.

Eroding wooden grave markers in the Victor Cemetery. Some names are familiar; others have forever vanished.

Today, no one can honestly answer the hypothesis of the wine glass of the Cripple Creek-Victor Mining District. Some believe the whole hypothesis to be nothing more than another monument to a fool, similar to Finn's Folly. But for others, the mystery is yet to be solved.

PERSONALITIES AND MINERS' LORE

Of the personalities, celebrities, and luminaries who either were born in the district or lived in the area for a time there are a great many stories. Some do stand out from the others and are worth mentioning because they characterized the over-all mood of the camp.

One of the most famous citizens to have come out of Victor was Lowell Thomas. Born in Woodington, Ohio, on April 6, 1892, Dr. Thomas moved his family into the "Greatest Gold Camp in the World" when Lowell was still a young boy. At the age of eight, a house on 6th Street became Lowell Thomas' first home in the district. The year was 1900, just before Theodore Roosevelt was to be mobbed in Victor for being an advocate for setting a standard price of gold at $20.67 per Troy ounce.

Lowell's father quickly became a highly respected doctor in the area. Because of his father's reputation, Lowell came in contact with many people. Dr. Thomas had to treat Lillian Russell for altitude sickness while she was on a stage tour through the area. Dr. Thomas adored the theater and met many celebrities of the stage through his medical practice. Anna Held, John Philip Sousa, and Lillian Russell were just a few of the celebrities that Dr. Thomas had to treat for one ailment or another while they were entertaining at the Victor Opera House.

Lowell Thomas started upon his event-filled career early. Before the age of twelve, he was delivering the papers of the *Victor Daily Record* in the tenderloin district of Victor and to residences in Cripple Creek as well as Goldfield. Even after he had learned what the ladies of the avenue did for a living he found respect for them and often listened to their tales and wisdom.

In 1904, a year of labor strife, Lowell Thomas was witness to the shooting of a man in the Victor Armory riot. Dr. Thomas and another man brought the wounded miner into Dr. Thomas' office. There, Lowell bravely aided the operation as his father saved the miner from a severe stomach wound.

Lowell Thomas worked in the mines, as had many of the residents of Victor. He had been an ore sorter, an assay carrier, a mucker, a trammer, and a driller during his boyhood and adolescence. At one time he ran carts out of the Empire Mine which had just struck a vein of ore assaying out at $36,000 per ton. Then he worked in the Tornado Stope, most commonly known for its ever-falling rocks. Riding assay for the Portland Mine was another job which gave Lowell Thomas a wide range of experience.

After attending the University of Denver, Lowell Thomas returned to Victor and again mucked in the Portland Mine. Shortly after his return, George Kyner of the *Victor Daily Record* offered him a position on the paper for $95 a month. By 1911, Lowell Thomas was editor of the *Victor Daily News*. From this position the world opened to him.

Lowell Thomas was always fond of his days in the Cripple Creek-Victor Mining District and told some intriguing stories about the area in his book, *Good Evening, Everybody*.

Many others of note resided in the area for a time. Groucho Marx and his brothers had been on tour with an acting company when they went broke in Victor. While staying at the Baltimore Hotel in Victor, the story goes, the wardrobe for the show was stolen and they were unable to go on with their tour. In order to work off some of their debt, Groucho Marx delivered groceries daily between Victor and Cripple Creek. The Marx Brothers were said not to have a name for their act then. Across the street from the Baltimore Hotel where they were staying was the Marx Clothing Store. Upon acquiring new costumes from the clothing establishment, they decided to take on the name of Marx — and thus became the Marx Brothers.

Jack Dempsey was working in the Portland Mine when he was discovered by trainer Morgan Williams. Dempsey's brother had been killed in the Portland in a mining accident. Still, Jack worked on in the mine. When Williams began to train Dempsey and realize his potential, he set up fights for him in the district. George Copeland fought Dempsey in the Cripple Creek Opera House. The bout lasted seven rounds, with Dempsey punching his way to a TKO.

Long before the streets of New York ever heard, "Hello, sucker!" Texas Guinan was playing organ in a church in Anaconda, just northeast of Victor. Marie Guinan was from Waco, Texas. She moved to Anaconda and became a teacher and Sunday organist. When she became the queen of the speakeasies of New York, her name had become Texas Guinan and her famous line simply echoed what used to be heard all around the Cripple Creek-Victor Mining District.

Texas Guinan had been one of the many adored personalities throughout the district. But when Bob Ford, the man who murdered Jesse James, came to Cripple Creek, Sheriff Wilson scared him right back out without so much as a meager greeting. Soapy Smith, an infamous Denver bunco man, did not receive such treatment. Instead, he made quite a bit of money in various con games from the salting of mines to highgrading to the fleecing of spectators at the Gillett bullfight of 1895.

Lon Chaney Sr. was an employee in the office of the power company in Cripple Creek. Tom Mix tended bar in his pre-Hollywood days on Bennett

Avenue in Cripple Creek. And Bernard Baruch was employed in several mines on Bull Hill.

The lore about the characters of the district is just as entertaining as the tales of the celebrities — such as the story of Ben Von Puhl. Ben Von Puhl was the brother of Sheriff Henry Von Puhl. Ben's profession was gambling. Ben was a professional: a man who played cards with a discerning eye and looked the part of the Old West gambler. But he had a kind heart. Ben would often beat a miner out of his daily earnings, then return the money when the miner pleaded that he had a wife and family to support. When Ben was well into his eighties, a casino in Las Vegas offered him $10,000 a year to have his presence for atmosphere in the establishment. Ben turned the offer down, stating that he was getting old and he wanted to die in Cripple Creek.

Then there was Jim Daily. Daily had been in a Midway saloon with A.E. Carlton when a mule-skinner suddenly threatened A.E.'s life. As the mule-skinner jerked his .45 out of his holster, Daily dove for the man. The gun went off. The bullet pierced Daily's jacket, but did not injure him. Daily beat up the mule-skinner and tossed him out of the saloon. From then on, Daily was a friend of Carlton's. A.E. even turned his head when Daily stole expensive mining equipment out of his idle mines.

The "Bull Hill Savages" were plentiful. All were eccentric in their own way. An old prospector named "Greasy" Miller was one of the Savages — or he could have been a composite of characters passed down in mining folklore. Greasy brought the first automobile into the district. After driving the two-cylinder Maxwell up the mountain, he parked it in a stable and never started it again. Many years later, the car was taken out of the stable and put in a car museum. The automobile had no more than fifty miles on its odometer.

The mining camp had been largely populated by Irishmen and Cornishmen. The Irish were called "Micks" and the Cornish were called "Cousin Jacks," (the wives of the Cornishmen were called "Cousin Jennys"). Along with the other laws that guided the operation of the mines, it was well-known that no Irishman was ever to work with a Cornishman. The historical war between Ireland and Cornwall had been brought to the American continent and given new life in the Cripple Creek-Victor Mining District.

Wild stories spread everywhere and added interest to the mythology of the Old West. The "Topic" was a parlour house where anything could happen — and did. Myers Avenue was heard about and respected as far away as Paris. But the stories of the mines and the miners were the most outlandish.

The one about how the Last Dollar Mine was discovered has its share of humor. It seems two Irishmen owned the claim of the Last Dollar. They had been digging and digging unsuccessfully for a very long time. Finally, the old bean barrel was about empty. So they started grasping at any hope.

They decided that upon their sworn last day on the claim they would dig wherever their old dog lay down to rest. Sure enough, the next day, in the bright sun, the dog found a place on the claim to sleep. No sooner was the dog comfortable when the two Irishmen kicked the dog off and began to dig. That evening, they uncovered the richest vein they had ever seen. Within a few days, they had sold the Last Dollar for $100,000 and trained off for Chicago. Unfortunately, in Chicago, the two Irishmen were taken by the first wild scheme that came their way. This story was probably started by a Cousin Jack.

The Wild Horse and Gleason mines above Midway. The discovery of the strike was credited to a horse which dragged a miner through the talc gold, covering him like butter.

The Wild Horse Mine has a good tale. One day its owner was trying to rope a rambunctious mustang. The days in the mine had been moving along moderately, but there was nothing better to do this day than rope horses for a new ore cart. The prospector lassoed the beautiful animal, but the mustang had more energy than he had. As the horse dragged the miner, his heels scratched into one of the richest gold-bearing talc veins ever found in the district. The miner immediately forgot the mustang. Taking off his boots, he scraped the gold talc off like cream out of a bowl.

And the stories continue.

There is the tale of how the Wright Brothers won an auto race which had taken place between Cripple Creek and Colorado Springs. The Wrights had stripped a car of all unnecessary weight and even cleared the back end so one could help balance the automobile on sharp corners by stretching out in the empty space. The Wrights won the race, but the only way they were able to stop their car was to roll it up a hill.

"Speed King" Wolfe was the owner of the first motorcycle in the district. His pride was increased every time he scared a mule or horse with the racing of the bike's engine. But he got a bigger thrill out of frightening pedestrians walking across Bennett Avenue.

Humane "Doc" Frank A. Hassenplug would rarely walk about town without eight or nine dachshunds at his sides. These dogs were often led down the street by one of Docs other assorted animals: a monkey, coyote, fox terrior, or turtle.

"Parson" Holmes was a hod carrier during the week and a street preacher on Saturday nights. He would try to scare the love of the Lord into those that frequented the saloons and sporting houses on Myers Avenue. One Saturday night while Parson Holmes was quoting religious adages while standing up pretty in his black frock coat in the back of his wagon, someone decided to spook his old horse. The wagon jerked forward as the horse began to run, and Parson Holmes was thrown forward into a sinful stretch of mud.

It was said that saloons and brokers offices matched office space foot for foot. Johnny Nolan's Manitou on Third and Bennett Avenue was a sight more elegant than most brokerage firms. His two-story gambling house and saloon had velvet curtains, Victorian wallpaper, roulette wheels, faro tables, and the highest-bidding poker games in the district. But Nolan was not a hardcore businessman. After letting the Salvation Army come into his gambling house during a severe rain, he let the organization have regular meetings in his place. Speaking briefly about the vices of alcohol and betting, the Salvation members would spot the biggest winners in the house and then pass the tambourine to the sound of the big drum.

One estimate of the district at its peak claimed that more than 50,000 people resided in the area. Of these people, only 4,000 were miners. At this time the monthly payroll for the miners was almost $650,000. And this did not include the managers' salaries.

Funeral processions out to the Cripple Creek cemetery below Mt. Pisgah were said to be common, the bands usually playing the deceased person's most cherished tune. Along the way, the blasting of the grave out of solid granite could be heard — most likely out of beat.

One day, George Keener, superintendent of the Mary McKinney Mine, announced to the miners on his particular shift that it was his birthday. Then he invited them all up after the shift to his office to celebrate. All the miners filed in gratefully as Keener handed them cigars. When they were inside, Keener calmly locked the office's door. Swiveling around, he pulled two .45's from his belt and told the miners to empty out their dinner buckets. Those with ore in their buckets were immediately fired. This scene must have happened before change rooms were established in the shaft houses.

The tenderloin section of Myers Avenue had actually been lit by red lights and the windows of the parlour houses colored with red shades.

Theodore Roosevelt rode one of the train cars through the tunnel which had gone under Squaw Mountain from the Gold Coin to the Economic Extraction Mill.

After Sam Strong, once owner of the rich Strong Mine, had become a millionaire, he acquired the reputation of being a touchy character about Victor. Strong had had a sour time of drinking one night and started roughing up the bartender of the Newport Saloon. A member of the Crumley gang took out a sawed-off shotgun and blasted Strong with buckshot through his right eye. As Strong lay dead on the saloon's floor, the killer proclaimed that Strong "had it comin' to 'im anyways."

People in the town of Anaconda used to laugh when someone came into a grocery store and bought food with stock from the Mary McKinney Mine. The people of Anaconda used to say that the grocers were being cheated. But the miners in the McKinney got back at the people of Anaconda. One day, a tenderfoot was testing out some new black powder in the Mary McKinney by using an electric shooter. Suddenly, the powder exploded. A huge hole was blown through the roof of the McKinney shaft house, and thousands of fragments of rocks showered down on the town of Anaconda. The strike produced eleven million dollars in gold ore.

Lieutenant R.S. Peary visited the Anchoria Leland Mine on Gold Hill. Some years later, his greatest claim to fame was an expedition to the North Pole.

Cripple Creek had a Pioneer All Star Tennis Club in its early days. One of the teams was called the Rah Rah Boys of the Nineties.

THE MILLING PROCESS

The gold tellurides found in the Cripple Creek-Victor Mining District are very rare throughout the rest of the world. Because this gold composition is so scarce, a special kind of ore-milling process has been developed over the years specifically for the Cripple Creek-Victor gold.

One of the first kinds of milling processes for gold tellurides was chlorination. This leaching method was widely used in the early 1890s. But it was soon found to be highly inefficient. Just before the turn of the century potassium cyanides were found to be much more productive.

Calaverite, one of the more abundant forms of the gold tellurides found in the district, contains 57.4 percent tellurium, 39.5 percent gold, and 3.1 percent silver. Sylvanite contains 62.1 percent tellurium, 24.5 gold, and 13.4 percent silver. Both of these major types of gold tellurides are generally found in the grayish-yellow breccia common to the Cripple Creek-Victor Mining District. The gold's appearance is more silverish in color than auric because the telluride compound has a strong resistance to oxidation. Iron pyrites and gold tellurides are similar in appearance and often fool the most seasoned of miners and prospectors. Hard quartz and granite usually encase the breccia bearing the gold tellurides, so it is obvious why the milling of the gold ore is a difficult and expensive process.

The use of cyanides in gold ore milling were first used in the year 1868 in Scotland. Many transformations occurred before the MacArthur-Forest process was introduced into the Cripple Creek-Victor Mining District in 1889. Chlorination milling was the most popular form of ore reduction at the time. This prominence of chlorination held until the late 1890s. The Woods Brothers built their huge Economic Extraction Mill just before the turn of the century, and they employed chlorination for the ore from their Gold Coin Mine. But as mining costs rose rapidly, more efficient ways of reducing the gold ore became necessary. Mill owners began to experiment more with potassium cyanides because the reduction rate of the gold ore had a much larger percentage of recovery than in the chlorination process, which meant less gold was lost in the mill slime and other reduction tailings.

The first cyanide reduction processes had four major steps. First, the ore was crushed and prepared for chemical additives. Second, a potassium cyanide slurry was mixed with the crushed ore. The third step dealt with the precipitation of the gold from the cyanide slurry through the addition of a zinc leaching solution. The fourth step concluded the process by heating the gold-zinc solution to a temperature at which the two metals separated. This, of course, is a highly simplified version of the process which requires a very large mill site and a vast array of heavy machinery.

The cyanide process went through many variations and modifications until it matured into a system of strict efficiency which has been used in the Carlton Mill since its construction. Originally built in 1951, the Carlton Mill stretches over three and one half acres in a gulch approximately half way between Victor and Cripple Creek on Highway 67. The mill has been improved throughout the years and is being upgraded once again because of the recent mining and milling agreement signed between Texasgulf and the Golden Cycle Corporation.

The mill can process 1,000 tons of ore each day. Reciprocating plate feeders and conveyor belts move the ore from each of the 16 fifty-ton-capacity storage bins to the mill's ore crushers. Classifiers take samples and rate the ore according to gold-bearing ratio contents as well as granular size. The gold values are rated then sent down either a high-grade or low-grade circuit. Again, conveyors and crushers are used to reduce the ore even further, and more samples are taken. From this point, both high- and low-grade circuits tend to become extremely intricate and complicated.

The following is a simplified version of the high-grade milling process which takes place in the Carlton Mill. The gold-bearing ore is thickened, then pumped into a holding agitator after passing through a very fine mesh. The concentrates are then passed through a filter, mixed with more solutions, then heated in a reactor. After being roasted, the gold-bearing calcines are mixed with a lime solution, then mixed with a calcium cyanide solution. The resulting mixture is agitated and thickened. After being clarified, a precipitate of zinc dust and lead nitrate is used to extract the gold from the solution. The resulting composition is roasted to separate the gold from the zinc.

One of the major variances between the low- and high-grade cyanide processes is the use of carbon-cyanide in the low-grade process. This carbon-cyanide step in the low-grade process was developed by the Carlton Mill, and increases the extraction of gold from the milling process.

The molten gold is taken from the roasters and poured into button- or cup-sized crucibles. Before the gold is shipped to the Denver Mint, it is roasted once more so that it is 99 percent pure.

The Carlton Mill has a 95 percent recovery factor in its milling of gold tellurides. The mill will resume daily operation in 1981 under the new mining and milling agreement signed by Texasgulf and the Golden Cycle Corporation. The Carlton Mill is expected to process most of the gold ore mined in the district.

EPILOGUE

There is an ever-present spirit pervading the Cripple Creek-Victor Mining District. It is a spirit that has more to its soul than gold. It is an aura that glistens with more than historic relevance. It is a feast of real dreams. It is the hope of endeavor and the reward of labor. And its physical embodiment is centered within and around the town of Victor.

Although Victor is well beyond the days of its first golden era, the spirit remains. More than gold had been at stake in the rivalry between the towns of Cripple Creek and Victor. More than worldwide recognition and central civic leadership motivated these two towns to strive for the prominent place in their era. The competition dealt directly with the conflict between realism and idealism.

The competition had started between the two towns as a race for the most beneficial position in the district, socially and monetarily. But as the years turned towards the Twentieth Century, the goals became much loftier. Supremacy for leadership in the growth of the state of Colorado became one of the leading factors. Then coming to the aid of the nation during economic panics entered the play. Soon the goal became a striving to be the most productive of human endeavors. This is where the town of Victor began to pull far ahead in the competition.

Cripple Creek was the center of financial exchange and social trend-setting for the district. Its image of wealth and high living gave the world the impression of a Gentleman's Gold Camp. All the commercial pomp and glamour which came out of the business sector on Bennett Avenue sported Cripple Creek as being the absolute center of the golden district. Financially, it was the district's focal point. However, the human side of the gold dream resided in Victor.

Victor was the major source of the gold tellurides. The town itself was called the City of Mines for some very specific reasons. The Independence, Strong, Ajax, Gold Coin, and Portland mines rose up right out of the town's actual city limits. These mines were more than mere holes in the earth to the people of Victor. These mines were their lives.

Most of the men who had lived in Victor were miners. These were the men who struggled daily with the earth and nature to remove the gold from the depths. These were the men who wrestled with the reality of the golden dream. Every day they had to live with the soft breccia and cold granite. Every day they had to battle with the mines' darkness, water, gases, and thin air. Each day was a fight with reality in order to maintain their dreams. Each day the miners' respect for nature and its rich dreams grew. This was the beginning of the fundamental difference between the miners of Victor and the businessmen of Cripple Creek.

The miners of Victor were more dependent on the mines than the lawyers, bankers, and merchants of Cripple Creek. For eight to ten hours a day, the miners toiled inside drifts hundreds of feet below the surface of the earth for a mere $3 a day. Mine owners living in Colorado Springs often accrued thousands of dollars each day simply because they had titles to the producing mines. Something more than the desire to survive kept the miners working. Something more than hope kept them from abandoning the mines. This spirit which let them escape mundane reality inspired them to form their own kinship of unity — a union of human idealism whose aura lives on in Victor.

The mine owners' fortunes were dependent upon their mines' production only until they could invest these monies in more stable businesses. Many mine owners gained their initial capital from the district's mines then moved on to other projects throughout Colorado and the rest of the nation. Neither their lives nor their fortunes were dedicated to the district. Most showed their disloyalty early in the history of the area by claiming their golden wealth and quickly establishing their homes in Colorado Springs. The miners did not have this choice.

Lawyers, bankers, mercantilists, and brokers were even more transient than the mine owners. Few of them remained in the area for a long period. Most came in to live off of the needs and misfortunes of others then left when other communities showed more promise. Only a scant few of the mine owners and other businessmen established residences in the district. Few had ever intended to stay long in the district in the first place. But the miners wanted the mines to remain productive so that they could build permanent homes, and settle down in an area which showed some stability.

To have a permanent home was uncommon for the miner. When the market bottomed out for silver, the miners had to find new homes wherever labor was needed. Creede or Leadville or Silverton would have been fine places to make their homes. But mining labor was no longer needed in these areas after the "Crime of 1873" and the silver crash of 1893 with the repeal of the Sherman Silver Purchase Act. The miners' communion with their special spirit was temporarily lost. Then gold was discovered on Battle Mountain.

The life of the miners was much different from the life of the mine owners. The miners' daily battle with the granite of the earth and their dream to build permanent homes with a steady source of income helped form the basic difference between them and the businessmen in Cripple Creek. This underlying difference nurtured the seed of rivalry between the towns of Victor and Cripple Creek. For the most part, the rich businessmen saw their dreams realized, while the miners were still thousands of mining drifts away from obtaining their aspirations. The town of Cripple Creek flowed with the fluctuations of golden wealth. The town of Victor constructed its foundations upon the hope of the gold ore still remaining in the ground.

The solidity of a union of labor established around ideals and dreams yet to be attained is the composition of the spirit which still resides in Victor. The surrounding hillsides are physical testament to the unity the miners felt during their strife throughout the historic era. The mine shafts, mine tailings, gallows frames, lonely shacks, and even the gold itself are monuments to the miners' labor and their unfulfilled goals. Long after the businessman's paper transactions have disintegrated into dust, the spirit of the miners' unity of dreams will remain. And Victor, Colorado, will always be a major source of that spirit.

The mark of the miner was destined to flourish decades after the mine owner's business disappeared. The City of Mines is not just the buildings which still stand in Victor. It is the people and the spirit which they share with the miners of the past eras. It is the Independence Mine, the Portlands, the Cresson Mine — and all the others known around the world for their wonders. All of this is the City of Mines. The legends such as the Wine Glass Hypothesis are part of the City of Mines. The history of the labor movement, the myths about the miners' personalities, and the lore about the miners' eccentric lives are all a pulsating part of the City of Mines.

The entire district was gold. The City of Mines possessed most of it. Cripple Creek was definitely the first home of a great many of the organizers of the district. The historical credit of Cripple Creek cannot be denied. But the favor that is due Victor, Colorado, has long since been put aside because it was not the romantic, commercial center that Cripple Creek had been. In the City of Mines, the human hand that toiled with the hard rock is still pulsating with the most cherished of unrealized dreams.

RECOMMENDED POINTS OF INTEREST

Victor

1) Victor Museum
2) Strong Mine
3) Independence Mine
4) Ajax Mine
5) Portland mines
6) Gold Coin Mine
7) Gold Coin Club House
8) Victor Cemetery

Carlton Mill (on Highway 67)

Cripple Creek

1) Cripple Creek Museum
2) Imperial Hotel
3) Palace Hotel
4) Old Homestead
5) Mollie Kathleen Mine (tours)
6) Mt. Pisgah
7) Cripple Creek Cemetery
8) Cripple Creek-Victor Narrow Gauge Railroad

Roads

1) US 24 and Highway 67 (Divide, Woodland Park, Manitou Springs, Colorado Springs, Denver)
2) Range View Road (well-graded dirt route between Victor and Cripple Creek)
3) High Road (well-graded dirt route between Victor and Cripple Creek)
4) Gold Camp Road (exhilirating back route between Colorado Springs and Goldfield, Victor, and Cripple Creek)
5) Shelf Road (excellent dirt route to Canon City)
6) Phantom Canyon (excellent dirt route to Canon City)
7) Cripple Creek-Florissant (back route to National Fossil Beds and town of Florissant)

A Brief Tour of Towns, Ghost Towns, Townsites, and Mines on the routes between Victor and Cripple Creek

1) Cameron — the first townsite encountered upon the Gold Camp and High roads. Known in 1892 as "Cripple Creek," Cameron was platted out as a townsite on October 24, 1894. The gold mines

Aspen-laden back road in the heart of the district.

in the area, such as the Cameron Mine, were moderate producers. At its height, the town of Cameron had a population of 500 and sported an amusement center for the district — Pinnacle Park — as its main attraction. Today, most of the townsite has been erased from physical history by mine tailings. Located at the mouth of Grassy Gulch, it is a cow pasture once more with only a few foundations of brick and stone left to indicate its past existence.

2) Pinnacle Park — this site is obscured by an abundance of aspen and pine, but the stone remnants of the animal cages can still be seen. South from the Cameron townsite, Pinnacle Park used to take up nearly a square mile of the hillside.

From High Road, the way north leads to Midway, Stratton, Cripple Creek, and Victor. The road south leads to Goldfield.

Taking the northerly route, an intersection of four roads is encountered. This is near the top of Tenderfoot Hill. Before the traveler's vision is the Hoosier Mine. Pursuing a northerly course, Hoosier Pass or Range View Road may be taken into Cripple Creek. Taking the furthestmost southwesterly route (a sharp left turn), the townsites of Stratton and Midway may be reached.

3) Stratton — founded by Winfield Scott Stratton to work his $50,000 per month Wine Glass Hypothesis atop Globe Hill. This townsite is to the right of the road on the southwestern Range View route. It has recently been sealed off to sightseers because of large-scale cyanide leaching operations on the mine dumps in the area. Established in 1900, Stratton was strictly a company town.

4) Midway — townsite to the north of Bull Hill where many railway routes and interurbans used to meet. Still standing, the Midway Saloon commands a majestic view of Pikes Peak to the east and backs a grand view of the Sangre de Cristo Range. The Wild Horse and Gleason mines shoot up above the townsite on Bull Hill.

5) Altman — from the center of Midway, on a route to the left of Bull Hill, the ghost town of Altman may be reached. Platted on September 25, 1893, its peak population — 2,000 — was reached in 1897. It was the general meeting place for the district's labor union. The road to Altman has been sealed off for the protection of the curious who are susceptible to sudden cave-ins which could leave one stranded some 1,800 feet below ground. Between Bull Hill and Bull Cliff, a safer view of this ghost town may be had from the ghost town of Independence.

6) Dyer — a junction point also known as Los Angeles because of its proximity to the Los Angeles Mine. This point is reached by continuing south on Range View Road from Midway, going past Windy Point, the Jo Dandy Mine, the John A. Logan Mine, the American Eagle Mines, the Orpha May Mine, Rubie Mine, and the Los Angeles Mine.

7) Independence — taking the left (east) route from the Dyer junction, the ghost town of Independence is easily found. This ghost town is very picturesque, with many of its buildings and gallows frames still standing. Between Altman (above the townsite) and Goldfield (below), Independence was platted on November 12, 1894. It is the soul of Battle Mountain. Independence is surrounded by such mines as the Vindicator, Golden Cycle, Last Chance, Deadwood, Delmonico, South Burns, and Sitting Bull.

The traveler may continue east to the Bull Hill Station, then take the Gold Camp-High Road back in a westerly direction to Goldfield. Or, the preferred route would be to return to Dyer — staying to the left — and resume traveling upon the Range View Road.

The western sector of Battle Mountain comes into view with an excellent overlook of Goldfield.

8) Goldfield — the peak population of this town was in 1901, when the residents numbered some 3,500. Today, 50 people live in the town. Platted on January 8, 1895, Goldfield was once the third largest town in the district and has suffered the greatest decline of the three major towns.

Continuing west, the wealthiest land of Nineteenth Century Colorado edges out all around the traveler. The Portland mines' gallows frames spring into view. Then the grand Independence Mine becomes visible. Right after the Independence Mine, the Strong Mine arises. These lands were easily worth more than $125 million when gold was worth only $20.67 per Troy ounce.

9) Victor — platted on November 6, 1893, this town had many interesting names. It was first called Lawrence. Then, at various times, certain sections of the town were named Hollywood, Portland Station, Strong's Camp, and Dutchtown. Today, the City of Mines is sufficient.

Range View Road turns into Highway 67 in Victor. Moving north on Highway 67, the route rounds Squaw Mountain and enters into Arequa Gulch.

The town of Goldfield was once chaotic with the bull pens set up by the Colorado militia during the violent strike of 1904.

10) Arequa — the oldest townsite in the district, founded in 1889. It was not until February 17, 1892, that it was platted. Sitting between Squaw Mountain and Beacon Hill, the town of Arequa was never very large.

11) Elkton — between Arequa and Eclipse gulches. This ghost town was never officially a town, but it sat close enough to the rich Elkton and Cresson mines to get on the map. From the year 1893, when it was settled, it took a decade to attain a decent-sized population.

12) Mound City — little if anything is left of this townsite opposite Raven and Beacon hills on their north side. The area had made a convenient site for ore mills because of its location near paying mines. Some of the foundations of these mills are still visible.

13) Anaconda — platted in 1894, this ghost town once had a population of 1,200. Between Gold and Raven hills, this town still has some stone and wood remnants for the adventurous traveler to see.

14) Cripple Creek — this town has also had many different names. After being platted on November 4, 1891, as Fremont, it possessed

such titles as Hayden Placer, Poverty Gulch, and "South Cripple Creek," with additional sectors of the town being known as Capitol Hill and Arcadia Heights.

Moving north on Highway 67, the Gold King, C.O.D., El Paso, Anchoria-Leland, and Proper mines become visible in Poverty Gulch and atop Gold Hill above Cripple Creek. The Mollie Kathleen Mine is the last major mine along Highway 67 driving north towards Tenderfoot Hill.

15) Gillett — "Bachelor's Town" was platted on January 18, 1894, and once was the site of Penrose's race track and many gambling houses. Eight hundred people once lived where only a few structures now stand. The stone foundations of St. Dimas' Chapel can still be seen.

At this point, it is still possible to turn right on High Road and circle back to Victor.

ENCOURAGING NOTE

Owning a piece of history seems to delight almost everyone. The entire Cripple Creek-Victor Mining District is a huge vault of vigorous history. The temptation to take a piece of the district home is almost overwhelming. Once one visits the area and acquires some basic knowledge of the history of the district, it becomes almost impossible to leave Cripple Creek and Victor without some kind of precious memento.

Acquiring a piece of history is encouraged. But, this encouragement does not include the destruction of history or private property. Almost all of the land in the district is owned by private individuals or corporations. They are quite tolerant of minor trespasses onto their property for the sake of a good picture or a better view. The dismantling of old structures, however, climbing upon the mines gallows frames, or exploring inside mines is politely discouraged. This is mostly for the visitor's safety. Many of the old mines are dangerous because of weak ground which could give way under the slightest weight and leave one stranded hundreds of feet below the surface. The mines and structures in the district are picturesque but should be viewed from a safe distance.

If the desire to own a piece of the Cripple Creek-Victor Mining District is irresistible, then take a stroll into the shops in the area and purchase an old bottle, an authentic mining certificate, or a small piece of gold ore. Many pieces of history can be bought in the area. Take a look in the shops in both Victor and Cripple Creek. The discoveries are intriguing. The history in the surrounding hills and mounts should be left so the visitor may look forward to returning for a much longer stay.

One of the few horizontal mine shafts in Poverty Gulch.

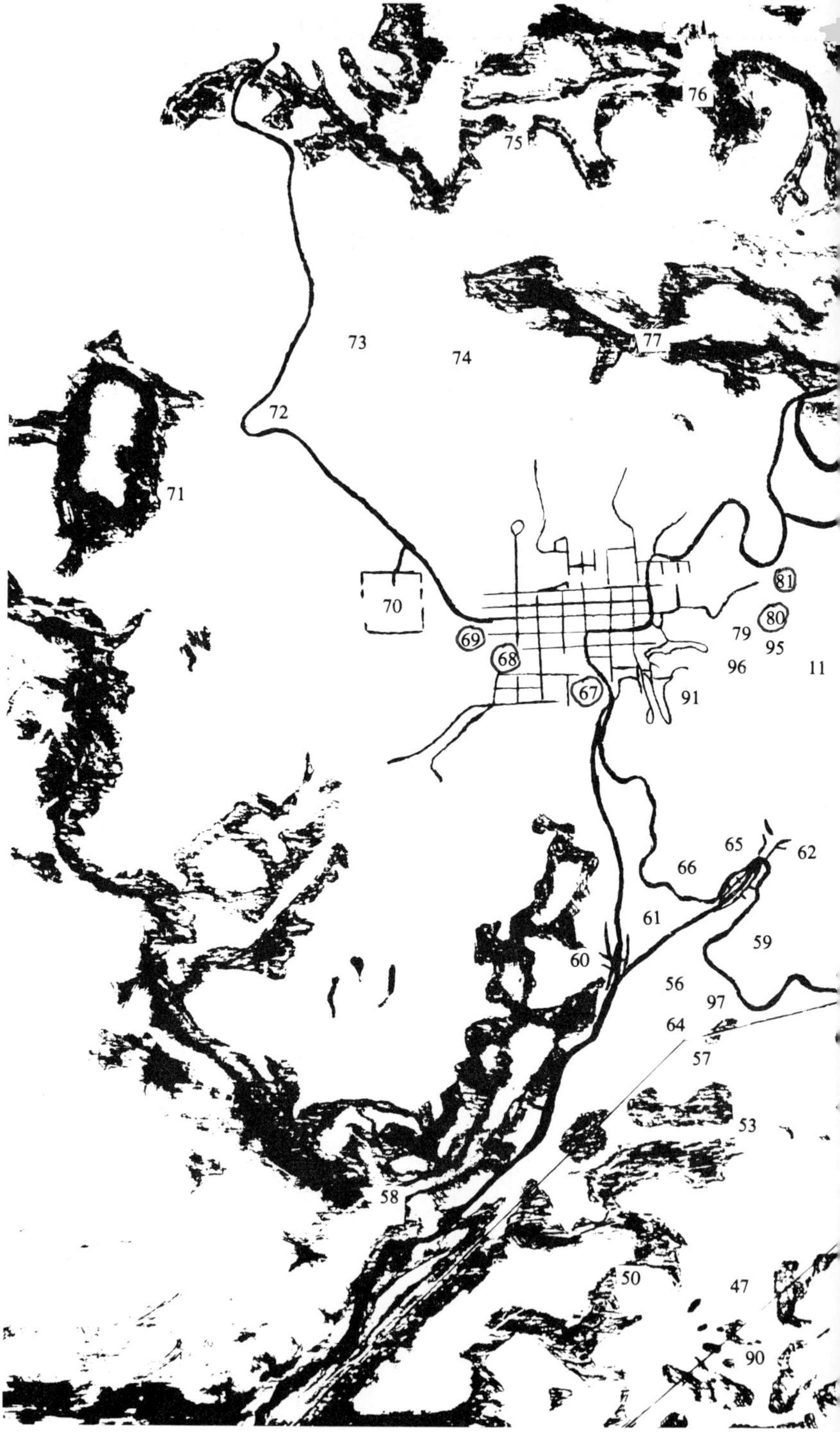
76
75
77
73
74
72
71
70
69
68
67
81
80
79
95
96
11
91
65
62
66
61
59
60
56
97
64
57
53
58
50
47
90

3
89
87
88
86
98
85
2
84
6
83
94
7
5
4
1
10
93
12
14
13
15
17
16
27
30
22
26
18
19
31
28
25
23
21
20
29
24
54
32
52
35
33
36
37
34
39
38
40
41
92
42
46
44
43
48

POINTS OF INTEREST ON CRIPPLE CREEK MINING DISTRICT MAP

1. Gold Camp Road - originally bed of Short Line Railroad (well-graded dirt road through Cheyenne Canyon)
2. High Road - cut-off from Highway 67 at Gillett: to Cameron, Range View Road, Goldfield, and Victor (well-graded dirt road)
3. Highway 67 - extension of U.S. 24 out of Divide, Colorado: originally bed of Midland Terminal Railroad (2-lane paved road)
4. Cameron (Town of), and Cameron Mine - ruins
5. Ruins of Pinnacle Park
6. Hoosier Pass/Hoosier Mine
7. Range View Road - well-graded dirt road through heart of mining area
8. Winfield (Town of) - ruins
9. Globe Hill
10. Iron Clad Hill
11. Gold Hill

    Between points 8 - 10: Lady Stith Claim - Plymouth Rock Shaft
12. Midway (Town of) - ghost town; also, Solomon Mine, Midway Saloon
13. Gleason and Wild Horse mines
14. Isabella Mine Group
15. Altman (Town of) - ghost town
16. Bull Hill
17. Windy Point - U.S. Geological Survey Point: Sangre de Cristos and Collegiate mountain ranges spread out before sightseer
18. Jo Dandy Mine
19. John A. Logan Mine
20. Cresson Mine
21. Raven Hill
22. American Eagle Mines
23. Orpha May Mine
24. Bluebird Mine dynamite bunkers
25. Independence (Town of) - ghost town
26. Deadwood Mine
27. Bull Cliffs
28. Findley Mine
29. Vindicator Mine
30. Victor Mine
31. Bull Hill Station
32. Golden Cycle Mine
33. Last Dollar Mine
34. Goldfield (Town of)
35. Battle Mountain
36. Portland mines (3 double compartment shafts)
37. Independence Mine

38. Strong Mine
39. Dillon Mine
40. Ajax Mine
41. Gold Coin Mine and Victor-Columbine Tunnel (on Diamond Avenue)
42. Hollywood (early section of Victor)
43. Victor (Town of) and Victor Museum
44. Lawrence (early section of Victor) and Mount Rosa Placer
45. Victor Cemetery
46. Phantom Canyon - to Skagway Dam and Canon City (partially paved and partially well-graded dirt road)
47. Carlton Tunnel
48. Straub Mountain
49. Squaw Mountain
50. Grouse Mountain
51. Economic Extraction Mill and Victor-Columbine Tunnel-foundations and ruins
52. Eclipse Gulch
53. Arequa Gulch - Townsite of Arequa
54. Elkton (Town of) - ruins, portions of ghost town
55. Elkton Mine
56. Beacon Hill
57. Roosevelt Tunnel
58. Gatch Park
59. Guyot Hill
60. Mound City (Town of) - ruins and foundations
61. Shelf Road - to Canon City through ruins and foundations of Anaconda (well-graded dirt road)
62. Anaconda (Town of) - ruins and foundations
63. Squaw Gulch
64. El Paso Mine
65. Anaconda Mine
66. Mary McKinney Mine
67. Cripple Creek (Town of) and Cripple Creek Museum
68. Myers Avenue (Cripple Creek)
69. Bennett Avenue (Cripple Creek) - Cripple Creek Museum at east end, and location of depot of Cripple Creek-Victor Narrow Gauge Railroad
70. Mount Pisgah Cemetery
71. Mount Pisgah
72. Road to National Fossil Beds Monument and Florissant (partially well-graded dirt road and partially paved road)
73. Red Mountain
74. Mineral Hill
75. Copper Mountain
76. Rhyolite Mountain
77. Carbonate Hill
78. Tenderfoot Hill

79. Poverty Gulch
80. Gold King Mine (Original El Paso Lode discovered by Bob Womack)
81. Mollie Kathleen Mine
82. Jerry Johnson Mine
83. W.P.H. Mine
84. Galena Hill
85. Beaver Creek Valley
86. Gillett (Town of)
87. Lincoln Strike
88. Penrose Race Track - close to sight of Wolfe's Bull Fighting Ring
89. St. Dimas of the Mountains Chapel - ruins
90. Wilson Creek
91. Old Fremont (one of the original sectors of Cripple Creek)
92. Victor Avenue - Victor Museum at south end
93. Pinnacle Mine
94. Grassy Gulch
95. C.O.D. Mine
96. Chicago and Cripple Creek Tunnel
97. Carlton Mill
98. Backview of Pikes Peak

Once a busy walkway into the Victor Mine near Altman.

BIBLIOGRAPHY

Feitz, Leland. *Cripple Creek! A Quick History of the World's Greatest Gold Camp.* Colorado Springs: Little London Press; 1979.

Looper, Kitty. *Road to Gold.* Colorado Springs: Little London Press; 1976.

Luna, Ricardo L. (publisher). *Cripple Creek, Illustrated.* Colorado Springs: El Napeste Publishing Co.; 1976. (originally Issued by the Passenger Department, Sante Fe Route, 1896).

Newton, Harry J. *Yellow Gold of Cripple Creek.* Denver: Nelson Publishing Co.; 1928.

Sargent and Rohrabacker (editors, writers, directors). *The Fortunes of a Decade.* Colorado Springs: Old Times Publishers Co. (reprint); 1979.

Sprague, Marshall. *Money Mountain.* Lincoln and London: University of Nebraska Press; 1979.

Taylor, Robert Guildford. *Cripple Creek.* University of Indiana; 1966.

Thomas, Lowell. *Good Evening Everybody.* New York: Avon Books; 1976.

Waters, Frank. *Midas of the Rockies.* Chicago: Sage Books, The Swallow Press, Inc.; 1949.